Integrated Electronic Payment Technologies for Smart Cities

AF694472

Don Graham

Integrated Electronic Payment Technologies for Smart Cities

Springer

Don Graham
Fort Pierce, FL, USA

ISBN 978-3-031-38224-6 ISBN 978-3-031-38222-2 (eBook)
https://doi.org/10.1007/978-3-031-38222-2

© The Editor(s) (if applicable) and The Author(s), under exclusive license to Springer Nature Switzerland AG 2023
This work is subject to copyright. All rights are solely and exclusively licensed by the Publisher, whether the whole or part of the material is concerned, specifically the rights of translation, reprinting, reuse of illustrations, recitation, broadcasting, reproduction on microfilms or in any other physical way, and transmission or information storage and retrieval, electronic adaptation, computer software, or by similar or dissimilar methodology now known or hereafter developed.
The use of general descriptive names, registered names, trademarks, service marks, etc. in this publication does not imply, even in the absence of a specific statement, that such names are exempt from the relevant protective laws and regulations and therefore free for general use.
The publisher, the authors, and the editors are safe to assume that the advice and information in this book are believed to be true and accurate at the date of publication. Neither the publisher nor the authors or the editors give a warranty, expressed or implied, with respect to the material contained herein or for any errors or omissions that may have been made. The publisher remains neutral with regard to jurisdictional claims in published maps and institutional affiliations.

This Springer imprint is published by the registered company Springer Nature Switzerland AG
The registered company address is: Gewerbestrasse 11, 6330 Cham, Switzerland

Paper in this product is recyclable.

“I would like to thank God and my family without whom I would not have been able to complete this work”

Preface

Over the past few decades, there have been several factors which have combined to create what we now call the "Smart City." One may ask the question: why now? Why are we developing all these new applications and systems for urban populations which we never had 60 years ago?

The answer lies in the two or three main developments which occurred in the latter half of the last century and early 2000s.

Firstly, there was the issue of climate change due to pollution and the negative impacts that it brought globally. Smog, nitrates, and other polluting chemicals were concentrated in large cities such as New York, London, Los Angeles, Chicago, and elsewhere. This pollution was causing major health crises for the urban dwellers and visitors. This created the need to address the congestion that causes this pollution and to make the changes to reduce it as soon as possible by as much as possible. Programs were put in place in many cities to measure pollutants from vehicles with diesel engines and to find ways to reduce such vehicles from entering the city limits. The flip side to these programs were to require more electric vehicles and "smart transportation" such as ride share and use of mass transit systems like subway and bus ridership. Such programs have been implemented in cities like London, Singapore, and others.

The changes that were made involved building new infrastructure which were the backbone of the "Smart City." This infrastructure involves Information and Communication Technology (ICT) to collect data from the residents and commuters to improve the services that government agencies offer to the city population.

Along with the need to reduce pollution, there came a new ability to handle large amounts of data. The era of "Big Data" came into being with the rise of the Internet and smartphones which were ubiquitous in the early 2000s. New methods to collect, store, and process large datasets of live data were designed, and a huge (petabytes) real-time data could be uploaded to the cloud.

These two factors resulted in a flurry of development activity, since we had both a need (reason) to address a congestion problem and a method of solution for the need.

This book provides an introduction to the systems and methods that have been used in implementing the smart city. The pertinent infrastructure, networks, and data collection are introduced. Technologies such as Internet of Things (IoT), Radio Frequency ID (RFID), Electric vehicles, Driverless shuttles, and data storage and processing are covered at the practitioner level. A rudimentary treatment of these technologies and their applications are discussed and examples of how they're used such as in traveler information message boards, smart streetlights that can provide information on where empty parking spots are in the city, and pedestrian monitoring cameras which can help reduce accidents is covered.

The focus of the book is the method of payment used by the population in an integrated transportation system in the smart city. Mass transit, ride-share, toll pricing for highways, and parking lots can all be connected on a shared system and paid on a single platform.

Blockchain is examined as a solution for integrated payments in a smart city. The book discusses blockchain and cryptocurrency at a beginner level. The fundamental concepts behind how blockchain works and why it would be suitable as a payment technology in a smart city environment is reviewed.

Chapter 1 provides an overview of the characteristics of the smart city, Chaps. 2, 3, 4, and 5 look at some payment systems for smart cities and the challenges or opportunities encountered when implementing them. Chapter 7 covers congestion pricing and using the analytic methods of mathematical optimization to reduce congestion in a smart city. A case study of Interstate 95 in the US State of Florida is presented as an example of application of the theoretical model developed in the chapter. The second half of the book covers smart city implementation in several US cities, and what specific technologies were used (i.e., electric vehicles, autonomous cars, cameras, smart street lights).

The last two chapters present some international applications of smart cities.

Fort Pierce, FL, USA Don Graham

Contents

About the Author

Don Graham's research interests include smart city analytics, AI/ML, congestion pricing, optimization, and supply chain systems.

The focus of his recent research work has been on smart cities, blockchain applications, forecasting for supply chain networks, and optimal algorithms for congestion pricing.

Don has presented his research at several national conferences such as the Transportation Research Board (TRB). He co-authored a book chapter in *Managing Supply Chain Risk and Vulnerability* (Springer 2009) and several articles on supply chain forecasting and optimization.

He previously served as a Data Science Professor at Northwestern University and taught at the Florida Institute of Technology in Transportation and Logistics for over 10 years, in addition he worked in industry at the Institute for Defense Analysis, Lockheed Martin, and AVIS.

Since starting his career as an Electrical Engineer in the Defense Department, he progressed professionally to the Transportation and Analytics field, where he has worked, taught, and researched for over 20 years, gaining a multidisciplinary background which he leverages across industries.

Don completed the BS degree at the University of Florida, MS degree in Operations Research at Columbia University, and the PhD degree in Transportation at the University of Central Florida.

Chapter 1
Introduction to Smart Cities

1.1 Introduction

Smart cities have become a major phenomenon over the last few decades. Due to rapid advancements in technology and the worldwide adoption of wireless communication and the Internet, urban communities have come to integrate and use the benefits that derive from these advances.

Wireless technologies allow a city to collect large amounts of data and the near real-time streaming of live transactions from the source to data storage and processing sites. Government leaders have the luxury of a huge amount of information from mass transit transactions to make decisions which increase revenue and improve the lives of their citizens.

Electric vehicles, connected vehicle technology, and driverless cars are all new transportation technologies that urban leaders can utilize to modernize and increase the efficiency of their city streets. Several cities have started the adoption of driverless shuttles to help with the congestion in the city. Some smart cities, such as Columbus, Ohio; Las Vegas, Nevada; and Detroit, Michigan, and Gainesville, Florida have implemented autonomous shuttles to supplement their public transportation systems. In many cases, these shuttles are partnerships between the city and an educational institution.

One of the immediate and apparent uses for collecting all the data is to use this information to warn and inform commuters of road conditions. Traffic can be dynamically routed based on the current demand on a roadway to minimize travel times as well as warn drivers of accident conditions up ahead.

In large cities, mass transit provides most of its population with transportation from home to work and back. Cities such as New York, Chicago, Los Angeles, and Atlanta have major public subway and bus systems which perform millions of transactions daily. This demand and ticket sales data can be (and is) used by city administrators to determine trends for the usage of transportation facilities. Travel trends are then used to forecast travel demand and modes of travel for each hour of the day.

© The Author(s), under exclusive license to Springer Nature Switzerland AG 2023
D. Graham, *Integrated Electronic Payment Technologies for Smart Cities*,
https://doi.org/10.1007/978-3-031-38222-2_1

One main impact of the data is in the planning function. City agencies can plan for and anticipate changes before they occur. Assets can be allocated to ensure that enough transportation facilities are available when needed. When a surplus of resources exists, the data can be used to distribute additional units, for example, trains and buses, to locations where it is needed most. For more information on smart city operations, see Saraju [1]. Data can be gleaned through Internet of Things (IoT) devices and transmitted by transmission devices and protocols to the cloud, where it is stored and processed. Once the data is processed by cloud applications, the appropriate commands can be sent back to the IoT device almost in real time. This allows for the remote operation of devices like cameras, street lighting, and other connected devices.

1.2 Smart City Overview

A smart city uses digital services to augment traditional infrastructure, increasing efficiency and coordination between public agencies and private companies. By using advanced technology built into the city's infrastructure, such as sensors that track traffic congestion or empty parking lots, public officials and developers can better understand the needs of their citizens and design cities to better suit their needs.

The urban population has increased over the last 30 years, a trend that is only expected to continue over the next 30 years to rise well above 60% of the total population. This increase brings congestion and major transportation problems for city administrators. These problems of congestion and climate change due to air pollution are driving the urgency to act to find a solution. At the same time, advancement in technology has shown potential to solve many of these problems if applied correctly. The combination of these factors has resulted in a concerted effort to develop technological applications for the smart city. Several cities have implemented these technologies and seen improvement in economic and sociological aspects of urban living standards.

Batty et al. [2] suggested that smart cities are made up of several layers that interact with one another. The emergent layer is composed of blockchain, IoT, and AI technologies that are enabling a new economy. They further suggested that smart cities focus on five dimensions: e-payments, digital identity, mobility, sustainable urban planning and buildings, and health.

In Barcelona, Spain, some smart city officials see the evolution of the smart city in three phases:

1. In the first phase, the smart city initiative focuses on technology such as sensors, cameras, and other technologies.
2. The next phase focuses on governance of the smart city.
3. In the third phase, the vision shifts to a customer-centric focus, putting the people first in any smart city plan.

1.3 Smart City History

Many people believe the smart city concept was developed in 1991 when the World Congress on Intelligent Transportation Systems was held in Toronto. The term "smart" referred to the interconnectivity of the city and its urban elements. A smart or connected city is ICT-enabled computer chips and sensors which are ubiquitous, permeating every aspect of urban life. It should go without saying, but it's not the sole focus of smart city development for governments to use ICT in the diffusion of services. Thanks to the development of social media, the focus has shifted toward a more inclusive and community-oriented approach that uses smart solutions (like wireless devices) to connect citizens to their cities and provide faster, easier access to vital services.

Although a wide range of technologies are implemented by smart cities globally, an analysis will show that a common thread is the use of sensors connected through communication networks which upload data to a central storage and processing facility. This data and network can be used to control devices such as a smart streetlight which can detect empty parking spaces nearby and share that information with connected users.

An analysis of the lack of financial innovation in smart city frameworks has led to developing a model that will incorporate electronic payments into smart cities. The model could also be modified and used to consider other financial services [1].

These are the aspects to be considered.

- *Smart Mobility*
    - Integrated transport that is ICT supported.
    - With smart mobility, there is less need for constant movement with the provision of prioritized clean and nonmotorized options. This is due to drivers having the information they need to go directly to their destination location with one trip (such as finding a parking spot with smart mobility) as opposed to making multiple trips to search for the same empty spot.
    - Following this, the public has improved access to relevant, real-time information. Time and costs are saved, commuting efficiency is improved, and CO_2 emissions are significantly reduced.
- *Smart Environment*
    - Resources are well maximized, ICT-enabled energy grids and metering.
    - Consequently, resource use is efficient through recycling.
- *Smart Governance*
    - Public, private, and civil organizations have access to integrated services and interactions enabling them to function as one organism.
    - Transparency and open data are central to smart cities.

- *Smart Economy*
  - Increased productivity through improved retail/commerce (e-business and e-commerce).
  - Focus on innovations and enhanced creativity facilitated by the presence of an inclusive society.
  - Enhanced value of data through the effective appropriation of data analytic tools and facilitating informed decision-making and creation of products and services following input, use, manipulation, and personalization of data.

1.4 What Are Smart City Technologies?

A smart city utilizes different types of technology and networks together with IoT to provide solutions for the public. Of the technologies, the IoT happens to be quite relevant [1].

The Internet of Things (IoT) is a network of multiple physical devices connected to the Internet to communicate and exchange data. IoT has a wide range of functions, and its adoption will improve smart cities greatly. This both gives room for improvements to be applied in each sector and boosts the economic benefits and improves the standard of living of citizens.

Several IoT devices utilize edge computing; edge computing ascertains data relevancy. Furthermore, a security system is adopted to monitor the exchange of data. In addition, this security system prevents intruder access to the IoT network of the city's data platform.

Smart cities also utilize other technologies in conjunction with IoT solutions. These technologies are:

- Artificial intelligence (AI)
- Cloud computing services
- Application programming interfaces (API)
- Mesh networks
- Machine-to-machine communications
- Machine learning
- Dashboards

1.5 What Are the Features of a Smart City?

Integrating the IoT, machine learning and automation will foster the implementation of smart city technologies for different applications.

There are many smart city features. Key features are energy conservation and environmental efficiencies, for example, streetlights that dim when the roads become empty.

Smart city initiatives can curb climate change and air pollution.

Apart from rendering essential services, smart cities provide their citizens with safety measures. For instance, they help in monitoring areas of crime. Moreover, they use sensors to warn in case of disastrous incidents such as landslides, floods, droughts, or hurricanes.

Smart cities can connect a variety of services to provide meaningful solutions for citizens. It works around several approaches or steps to optimize the lives of its citizens and facilitate economic growth using IoT solutions and other technologies. This strategy lowers costs, enhances sustainability, and streamlines factors like energy distribution.

Transactions are an essential part of a smart city. Mothobi et al. [3] discuss how infrastructure deficiencies and the adoption of mobile money in Sub-Saharan Africa impact integrated payment. The vision of smart city centers around providing services that bring economic growth and a comfortable life to its citizens. The services the government provides form a significant part of the economic activities of a smart city, which includes salary payments, tax collection, and business procurement.

1.6 Smart City Data

As previously discussed, technology is the basis for the smart city. In fact, the "smart" in smart city refers to the information gleaned from the technology in the smart city. This technology provides large amounts of data, which can lead to informed decision-making. The data retrieved from cameras, sensors, and mass transit payment systems are all actionable data which administrators can use to improve the lives of the citizens and the economy of the smart city.

We will look at what technologies are used in the smart city and the implications of using this digital technology.

1.7 Data Flow

When we analyze the smart city, one question can be asked, "how does the data get from the source to the destination or from the transaction point to the point of storage?"

The system can be divided into four levels or layers with many technological applications [4].

1. Infrastructure
2. Network
3. Data exchange/processing
4. Storage

The telecommunications infrastructure includes wired and wireless physical components which form the backbone of digital connectivity (see United Nations report on Connectivity May 2021, https://www.un.org/ohrlls/sites/www.un.org.ohrlls/files/21-00606_1e_ldc-digital_connectivity-rpt_e.pdf). This includes the cables and wireless 2G, 3G, 4G, and 5G capable components which are available to support data transmission.

The network level refers to the type of communication protocols which connect the devices in the network. The data transmitted with compatible devices and protocols can then be exchanged on the world wide web or via a cloud system.

The final level is data generation which produces information from the data source to the processing facility, where it is stored and analyzed.

Data can be generated from cameras, sensors, RFID devices, and even smartphones. When all these devices are connected through a network and transmit data into a central storage/processing facility, this is commonly referred to as the "Internet of Things (IoT)."

The connection of different devices into a central facility also allows for the control of these devices by intelligent applications or humans. Thus, for example, a smart streetlight can send a signal to illuminate or dim at specified intervals depending on the ambient light or the time of day.

Other examples include traffic lights that can change depending on the level of traffic flow sensed along a corridor. The sensor information along the corridor detects the level of occupancy at certain points on the road. It streams that data into a cloud application which can then perform calculations and apply algorithms to determine how much green time and red time should be allocated to the traffic lights at the intersection. In this way, the smart city can monitor traffic and reduce congestion for improved commutes and reduced travel time in the city.

The use of aggregate data in this way is one of the hallmarks of the smart city, as it can be applied to improve the lifestyle of the citizens and increase revenue or cut costs for public administration.

In many ways, one can visualize a cycle of information flow in the smart city, which iteratively improves living conditions and economic activity in the smart city.

The devices in the network generate huge amounts of data, typically in a cloud.

This data, often containing petabytes of information, is commonly referred to as "Big Data."

Benefits derived from the collection and possession of "big data" are many, and the administrators can improve many services such as transportation, utility, energy, and even waste management services. By analyzing the data, trends arise, leading to new services that previously were not provided due to the lack of knowledge that such services were needed.

The development of the smart city has revealed socioeconomic, sociopolitical, and sociotechnological inter-relationships, which show a synergy between the different groups. These groups, when acting together, can create new wealth for the citizenry and a better standard of living within urban environments.

The growth of the smart city has transformed the urban community's citizens from just a consumer of data to both a consumer of data and a generator of data. By doing so, the information provided by the citizens' use of cellular phones, cable TV, and remote meters in residences leads to a win-win situation for private individuals and the administrators who analyze the data for opportunities to improve economically, politically, and socially (Figs. 1.1, 1.2 and 1.3).

In the above graphic, the first figure shows the old model where citizens would only consume and pay for data. With the ever-expanding growth of the smart city, citizens are both consumers and creators of data, presenting a paradigm shift in how business is conducted.

This shift is beneficial to the consumer since the data they produce is used by their smart city to tailor services for them, as individuals and as a collective.

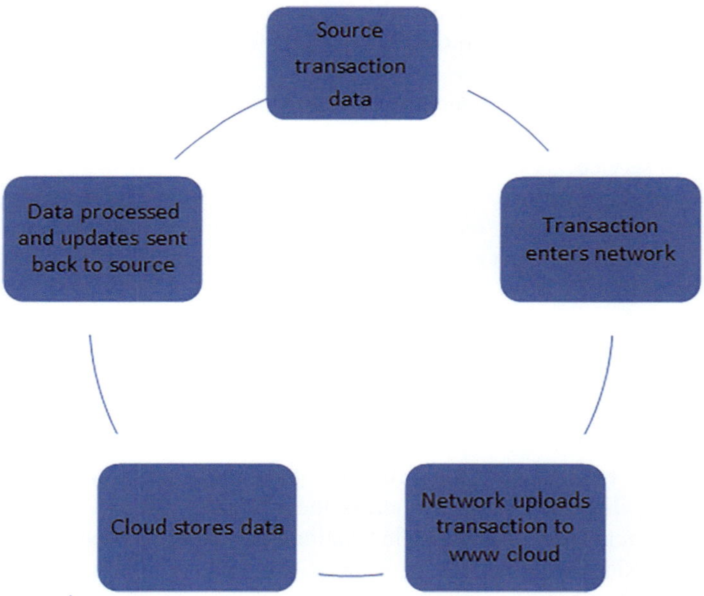

Fig. 1.1 Cycle of information and data flow in a smart city

Fig. 1.2 Citizens as pure consumers of data

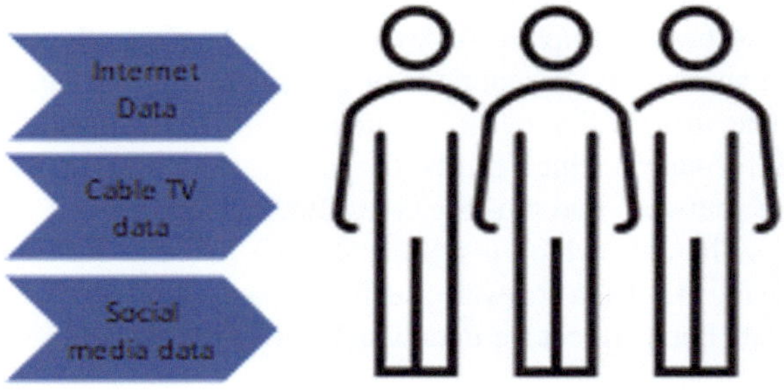

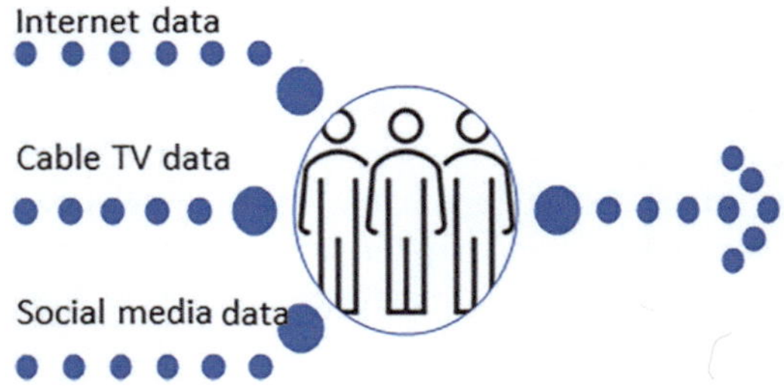

Fig. 1.3 Citizens as consumers and generators of data in Smart City

The progression of the smart city has been promoted by major stakeholders who can benefit economically from its wide adoption. Companies that manufacture infrastructure components (wired and wireless) stand to benefit from smart city implementation. In addition, many utility and telecommunication companies will see significant revenue through the wide installation of their components, such as smart meters.

The data generated is fundamental. Government leaders in a smart city will also rely on digital data analysis from artificial intelligence and data science firms who provide such service.

Because of these factors, the smart city is likely to grow exponentially over the next several decades as new technology makes the implementation faster and less expensive. The social impact is also a driving force since people see the benefits and improvements in their lives.

Smart cities have caused significant disruption in several industries. In the energy sector, smart cities have impacted the generation, transmission, and distribution of power. Solar cells placed on private residences have changed the centralized nature of electricity distribution since residences can not only consume power but generate their own power and sell surplus power back to the electric grid.

Wind energy has allowed major energy companies to source their power from geographically dispersed locations across the USA and abroad. This is in stark contrast to the generation from a single location, such as an oil-producing state near the Gulf of Mexico.

Because of the newly decentralized nature of electricity production, it can be a challenge to coordinate the stakeholders and processes within the smart city context. We will look at these challenges and how to address them later in this chapter.

Along with the energy sector, the telecommunication sector of the smart city has also been disrupted. As we mentioned before, the citizen has now become both a consumer and a generator of data, resulting in huge amounts of data for analysis. The availability of this data is good because of the trends and other valuable information which can be derived from it. This amount of data can cause delays in the network and must be controlled.

The data which is collected from IoT and stored in the cloud can be analyzed using Big Data software, such as Apache Hadoop. Telecom companies not only sell data but also receive data which can bring on new issues such as privacy issues and new regulatory control of what data can be reported or used and how it is stored and used. These new considerations will be addressed in the later sections of this book.

When administrators decide to implement a smart energy system, several important policy decisions need to be considered. These will typically be related to economic, technological, and societal impacts.

1.8 Smart City Concepts

Smart cities are technologically advanced urban areas that use different modes of technology (information and communication technologies) like voice recognition and sensors to gather specific information [5]. Data is collected from citizens, devices, buildings, and assets to understand and manage traffic, power plants, schools, and other systems. Smart cities are about more than how they use technology, and they are smart in the way that they monitor, analyze, plan, and govern the city itself.

The smart city integrates information and communication technology. This integrates with the citizens, creating an efficient ecosystem for urban development. The technology enables interaction with the government and to monitor what is happening in the city. A smart city can offer its citizens a better experience because it is prepared to react to emerging issues. However, there is no single definition of what it means for a city to be "smart."

1.8.1 Characteristics of a Smart City

A smart city applies technologies to improve the effectiveness and efficiency of services, such as the delivery of electricity, water supply, and management of traffic congestion. It also improves citizen satisfaction, economic development, and social equity. These technologies improve city performance by increasing efficiency and lowering the cost of delivering services [1].

A smart city uses information technologies owing to the following:

1. Technology is a boon to the global economy. Utilizing this technology on infrastructure supports a strong and healthy city. These are the same goals that cities have been working toward for decades, but technology just might be the missing piece of the puzzle.
2. As it becomes more difficult "to know what you don't know," smart cities are adopting an intelligence-driven approach. They are learning, adapting, and innovating in the way they gather information and analyze data.
3. They are getting smarter every day. Cities' knowledge and cognitive competence are increasingly effective with digital communication networks and ubiquitous

sensors and tags. The mix of telecommunication, intelligence, and senses makes a city intelligent.

These forms of intelligence are as follows.

1. Empowerment intelligence
   In Stockholm, Kista Science City (www.kista.com) is an open platform for innovation. In Hong Kong, the Cyberport Zone is a similar facility. In Melbourne, a different open platform has been set up.
2. Instrumentation intelligence
   The use of data collection in smart cities is both controversial and increasingly prominent. There are concerns about increased surveillance in these cities, but examples like Amsterdam show that data collection can be beneficial. Amsterdam has implemented this using instrumentation intelligence to facilitate the following:

   - Applications to build a smarter planet.
   - Creation of an environmentally friendly planet.
   - Smart energy meters are being installed in many homes across the country to provide people with information about their energy usage and help them minimize their energy consumption.

1.9 The Idea of an Integrated Smart City

The integrated smart city is one in which the administration uses information technology to improve the lives of its citizens. Smart cities became a practical idea once technology matured to the point it is today. Large amounts of data can be stored and processed in very small amounts of time and, in some cases, in real time.

This ability to process massive amounts of data, sometimes referred to as Big Data, has allowed large volumes of people to process transactions.

A smart city uses data as a central and prominent feature to optimize its resources, energy transportation, and economy.

The smart city is built around its citizens and so it must encourage the development and usage of data services by its residents. Optimizing transportation and energy networks is also required from its citizens.

So, a smart city is, above all, a city that uses existing data to allocate resources optimally for the benefit of the entire population in the city [5].

Smart city planning requires a new way of thinking. A smart city planner must be flexible and willing to adapt to agile project planning techniques in order to maximize the impact of the data used and gain the most positive results from the application of data-driven decisions.

This shift in thinking can sometimes cause problems for those who are used to traditional planning methods and result in obstacles that must be overcome to succeed. However, despite these obstacles and risks, achieving the maximum benefits from a smart city greatly outweighs the challenges.

1.10 Smart Contracts

A smart city must integrate its financial networks to provide a smooth experience for the transportation modes and energy payment systems.

Blockchain is one emerging payment technology that has seen exceptional growth over the last 5 years. In this book, we will introduce the concept of blockchain and cryptocurrency as a payment method in a smart city. Nakomoto [6] introduces Bitcoin as a cryptocurrency used with the blockchain. A central feature of blockchain is the concept of a smart contract.

The concept of the smart contract was developed more than 20 years ago in a paper by Szabo [7].

One implementation of smart contracts is within the Ethereum blockchain. A smart contract is a conditional transfer, an application-specific validation system for blockchain transfers.

Smart contracts can transfer assets besides cryptocurrency. The smart contract specifies policies for the transfer of assets over the blockchain. Smart contracts represent business logic that involves a messaging system to be executed conditionally over the decentralized network.

When funds are transferred, it primarily purchases products, services, or items of value. These transfers do not consider the business policies or rules of the parties involved. In the real world, there are many transactions where business policies must be taken into consideration.

Example transactions are automobile loan applications or mortgage loans, where the loan is only approved if the applicant has a certain credit score and income. These types of rules can be handled by a smart contract as they require more processing than the simple transfer of funds.

Smart contracts contain program code to execute the business rules. Since the smart contract is deployed on the blockchain, it cannot be changed. However, a new smart contract can be implemented in place of the previous one.

References

1. Saraju, P. M., et al. (2016). Everything you wanted to know about Smart Cities. *IEEE Consumer Electronics Magazine, 5*(3), 60–70. https://doi.org/10.1109/mce.2016.2556879
2. Batty, M., Axhausen, K. W., Giannotti, F., et al. (2012). Smart cities of the future. *The European Physical Journal Special Topics, 214*, 481–518.
3. Mothobi, O., & Grzybowski, L. (2017). Infrastructure deficiencies and adoption of mobile money in Sub-Saharan Africa. *Information Economics and Policy, 40*, 71–79. https://www.sciencedirect.com/science/article/abs/pii/S0167624516301342
4. Connectivity in the Least Developed Countries. *Status report 2021*. Retrieved from the world wide web May 2023. https://www.un.org/ohrlls/sites/www.un.org.ohrlls/files/21-00606_1e_ldc-digital_connectivity-rpt_e.pdf

5. Novotný, R., Kuchta, R., & Kadlec, J. (2014). Smart City concept, applications and services. *Telecommunications System Management, 3*(2). https://doi.org/10.4172/2167-0919.1000117
6. Nakamoto, S. (2008). *Bitcoin: A peer-to-peer electronic cash system*. https://git.dhimmel.com/bitcoinwhitepaper/v/a5f36b332cb6a5fa9e701886f376ac1ac2946d07/
7. Szabo, N. (1994). *Nick Szabo's papers and concise tutorials*. https://www.fon.hum.uva.nl/rob/Courses/Informationinspeech/CDROM/Literature/LOTwinterschool2006/szabo.best.vwh.net/smart.contracts.html

Chapter 2
Payment Technologies in the Smart City Context

2.1 Introduction

Payment technologies in a smart city depend on data and the transfer of data from a debtor to a creditor. Payments for goods or services have had a long history since the beginning of commerce.

Barter was one method used when people began exchanging goods for other goods or services for services without the need for a standard currency. Paper money was introduced as currency for the first time around 700 B.C. by the Chinese. Paper currency has continued to be used globally for currency from that time until the present.

Over the past several decades, as financial institutions became more technologically advanced, credit cards and debit cards were introduced for the convenience of the customer and improved digital recordkeeping.

As the Internet evolved, transferring money digitally became a reality. Companies such as PayPal, Venmo, Cash App, Zelle, Square, and WeChat saw wide acceptance from millions of customers in everyday usage.

Fully online banks such as Ally are also able to provide deposit accounts and make auto loans as well as mortgage loans.

In this chapter, we will introduce and examine these types of payments and their use in smart cities and future trends. We will also introduce the new area of Fintech and concepts such as digital wallets and mobile payments where relevant.

At each step, whenever an innovation in payment occurs, it is due to the new method being more efficient and /or easier to use. For instance, cards are seen to be more convenient to use by millions of customers than cash due to the ease of having a single card with monetary value rather than carrying large amounts of paper bills which can easily be lost or stolen.

One of the key features of the new digital payment revolution is that many of the innovations are spearheaded by technology firms as opposed to banks or financial

© The Author(s), under exclusive license to Springer Nature Switzerland AG 2023

D. Graham, *Integrated Electronic Payment Technologies for Smart Cities*,
https://doi.org/10.1007/978-3-031-38222-2_2

institutions. Companies like Google, Apple, and Facebook have developed their own payment systems for transferring money for goods or services online.

They have been able to take the lead in digital payments due to their level of technical expertise and fewer regulatory systems than financial institutions.

2.1.1 Payment Overview

Prior to the digital payment revolution, traditional payments involved a bank with a central clearinghouse. The four modes of payment in that system involved:

1. Cash transaction
2. Electronic fund transfer (EFT)
3. Cards (credit/debit cards)
4. Financial instruments

Cash transactions typically involve a small amount of funds transferred face-to-face. Paper bills pay for goods or services between private parties or from a private party to a retail establishment in person.

Cheques are a form of payment where one entity (private party or business) presents an order allowing a withdrawal of funds from the initiating party's bank account. The receiving party can take that order (the cheque) to the payer's bank, where the bank verifies the information on the cheque and pays the receiving party the amount of the face value of the cheque.

Card transactions involve debit or credit cards. A debit card allows the card user to pay for services through an electronic fund transfer at a merchant or retail facility. The debit card is tied directly to the holder's bank account and balance. The holder may not use their debit card to charge or pay more than the balance in their account unless agreeing to a fee with their bank.

Credit cards can be used to pay for goods or services up to the credit limit for which the holder is pre-approved for. Any balance over the pre-approved amount will incur a fee if this is approved.

Electronic fund transfer can be done with either a credit or debit card, where funds are transferred from the payer's account into the recipient's account remotely (typically using a computer). Electronic fund transfer can be initiated by either the payer or payee.

Banks and financial institutions send and receive payment orders on a network that can transmit electronic messages. These messages contain the orders that the banks must perform to settle the electronic transaction. One such messaging system is the SWIFT system.

Clearing houses or settlement banks are intermediaries between the member banks and provide settlement services between financial institutions.

2.1.2 Digital Revolution in Finance

To fully understand the changes that are taking place in the financial world, we must understand the terms involved. Digitization involves transforming a physical product or service into a digital product or service. For instance, a book that can be bought in hard copy in a brick-and-mortar store (such as Barnes and Noble) is converted into a digital form (i.e., pdf) and sold online on a webstore. The other concept of digitalization is the more general impact that digitization has on the system.

We can better appreciate the events of the present state and environment we live in by reviewing some of the historical advancements in digital technology applications.

The first automated teller machine (ATM) was developed and used by Barclays Bank in 1967 in England.

The first electronic stock market system was NASDAQ which opened in 1971 as merely a stock quotation system. This later evolved into a trading system where traders could buy and sell stocks electronically.

SWIFT messages were first used in 1975 on banks and financial institutions.

Internet banking was first introduced in the early 1980s. After a slow start, Wells Fargo was the first bank to offer a full banking service in 1995.

Mobile payment systems were an innovation that started in 1999 with the download and payment for mobile phone ringtones on Samsung phones.

These innovations started an influx of new technology improvements and additions when the Internet gained widespread usage and accessibility over the next 10–15 years. The financial sector has continued to advance by adding wealth advising services, and trading apps such as Robinhood have become household names.

2.2 Payment Technology Innovations

Payment technology innovations can be subdivided into those that are centralized and those that have decentralized infrastructure.

Digital wallets (e.g., PayPal) are innovations that do not require a decentralized system infrastructure.

Similarly, credit card systems have a centralized model for processing.

The payment innovations which rely on decentralized infrastructure include blockchain, cryptocurrency, and smart contracts.

2.3 Digital Wallets

A digital wallet is a digital payment transfer system that uses a private third-party bank to transfer funds from the sender to the receiver. The digital wallet utilizes existing automated clearing house (ACH) infrastructure. The digital wallet has the advantage of only requiring the input of the sender's (or receiver's) banking information one time to send or receive money multiple times [1, 2].

An example of a digital wallet is PayPal. A transaction can occur on a digital wallet (i.e., PayPal) when:

(a) Both the sender and receiver have provided their phone number, email address, or a username.
(b) Both the sender and receiver have linked their bank accounts to the wallet platform (website).
(c) The sender has available funds in their wallet account or bank account.

When the sender wishes to pay someone on the platform, they initiate the request by signing onto it with their username, phone, or email. The sender then enters the information (phone, email, or username) of the member they want to pay as well as the amount to pay them. The payment is then sent through the platform to transfer funds to the payee.

Once the payment is made, the receiver sees the money in their wallet account on the platform. At that point they may withdraw the funds and store it in their personal account.

In the background, the digital wallet will connect to the sender's bank account as the source to withdraw the funds. The members of the platform can "put money into their digital wallet" simply by transferring funds from their bank account to the wallet account on the digital platform.

This transfer of funds is done by initiating an ACH request from their bank to send money to the platform's (PayPal) bank account. Although digital platforms (such as PayPal) are not banks, they maintain an account at a bank where these funds are held.

Once funds are in the digital wallet, the sender can pay anyone who is on the platform by using their phone, email, or username. This results in a transfer of funds from the sender's wallet account to the receiver's wallet account instantaneously. Here is where the efficiency improvement of the digital wallet is realized. The funds are immediately available in the receiver's wallet account for use in payment or withdrawal by ACH.

Digital wallet platforms make money by charging a fee for transferring funds from the platform to traditional bank accounts. Typically, this is free for most customers but there is a fee for merchants or those who want premium services.

Since the digital platform is all software, they do not incur costs for building new infrastructure. Customers also benefit from not having to recall their bank routing and account numbers every time they do a transfer or make a payment.

With the typical digital wallet, the digital platform places the members' funds into an interest-free account where the funds are not used for any purpose until the member makes a request to send or receive money. This is a disadvantage for most members since they have no incentive to keep money in their digital wallet.

One digital wallet platform—Alipay in China—addressed this problem by placing the funds in the members' digital wallet in an interest-bearing account [2].

Thus, the members of the platform are incentivized to keep funds in their digital wallets with the expectation that the funds will grow over time. Alipay also increased its services to include investments, creating a digital wallet ecosystem offering multiple services to its members to increase their wealth.

Social networking platforms have also started their own payment systems. Facebook Pay is one payment that uses a digital wallet in a customized manner to allow users to pay for services on the Facebook platform. Due to the ease of use and the user not having to leave the social network site to pay another Facebook friend, many users are drawn to use this payment method on Facebook Marketplace.

Another social network type example is Venmo, a payment app that uses digital wallet technology (similar to PayPal) for payments. Venmo has added a social dimension to its payments platform by adding a payment as a status update on the social site.

Other payment systems with digital wallets are:

Stripe: useful for online sales and ecommerce.
WeChat: based in China and implements payments in the social network platform.

2.4 Decentralized Payment Infrastructure

As mentioned earlier in the chapter, a second type of payment technology relies on a decentralized infrastructure.

The decentralized payment technologies do not require a central bank or financial institution to process the transfer of funds or payments from sender to receiver [6].

The most common type of decentralized payment technology is blockchain and cryptocurrency.

Blockchain was developed initially near the end of the 1900s but really became a major player within the last 15 years. Blockchain is a decentralized ledger system that keeps track of transactions electronically.

Blockchain utilizes encrypted messages to perform transactions and update the network ledger; this process improves the security of the transaction and makes fraudulent transactions much less likely to happen. Any change to one node in the blockchain network will be seen by all the nodes, which will cause an alert to unauthorized activity.

For a transaction to be completed, it must be validated by decrypting the message and mining the transaction. Once it is validated, the entire network is updated simultaneously with the new balances of the ledger.

Blockchain utilizes cryptocurrency, such as Bitcoin, to perform the transaction and increase or decrease each account's value. Due to the increased popularity of Bitcoin over the last 5 years, the cost of purchasing Bitcoin has increased by more than 30 times [6].

2.5 Methods of Payment

In many smart city roadways, the managed lanes utilize apps for the payments of Express Lanes and HOT (high occupancy toll) lanes. These lanes are dynamically priced based on the level of traffic flow. SunPass in Florida, E-ZPass in mid-Atlantic regions of the USA, and Peach Pass in Georgia have utilized apps for smartphones that will accept payments for the ability to drive on the managed lanes. Belgavi [3] discusses digital payments within the smart city.

Credit and debit cards are used as the payment methods on these apps for the service to use the managed lanes.

Credit cards began with the Diners Club card. This business model is one where the bank extends an instant revolving line of credit to the customer and pays the merchant immediately. The customer is billed for the purchase later, while the banks assume the payment risk. This payment information is delivered over a message network while charging the merchant a fee for processing and making this guarantee on behalf of the customer.

The BankAmericard from Bank of America was then introduced and later became known as VISA. Competition in the credit card industry resulted in Master changing its credit card, which later became Mastercard.

When customers use their credit card for a purchase, they will swipe, insert, or tap their card on a credit card terminal to pay. The information of their account and the purchase amount are then routed over a network to the issuing bank instantaneously. This bank then decides whether the account has the available line of credit to pay this bill amount and does an analysis of whether this may be a fraudulent transaction. Once payment is approved by the issuing bank, the message is sent over the network to the merchant of the approval status and the bill is considered paid.

All the banks involved and the network charge a fee to process the transaction and make payment, which could be in the range of 1–2.5% of the cost of the amount charged.

The money is then transferred from the issuing bank to the merchant's bank, and the customer is billed for the purchase.

Within the context of a smart city, many transportation services are paid for with a credit or debit card. Ride-sharing companies such as Lyft and Uber can be paid using a credit card on their app. Mass transit also has traditionally utilized website payments with a credit or debit card. Based on this model, we see that most payments within the smart city will use a credit card or mobile payment system for the transportation services they provide.

2.6 Credit Cards: Advantages

A main advantage of credit cards is speed of processing. The merchant is paid almost immediately if approved. Another advantage is convenience to the customer. They can pay without carrying large amounts of cash with them.

The next advantage is that the customer is not usually responsible for the transaction in the event of fraud, and the bank typically bears the risk involved in the payment.

Dispute resolution mechanisms are in place should the credit card purchase not go as planned and the customer does not receive the expected goods or services.

2.7 Credit Cards: Disadvantages

Disadvantages of credit cards include:

- Credit card fees mean the customer pays more than the cost of goods sold.
- The customer pays an interest rate to the bank issuing the card, which adds to the total bill.
- For the banks, credit cards have significant security risks which in turn lead to higher fees.
- Terminal fees mean some merchants may not accept credit cards.

Despite these disadvantages, due to the ubiquitous nature of credit cards in the population, the smart city has seen high usage of credit cards for payment.

2.8 Integrated Payments

One of the roles of the smart city is to integrate multiple service providers onto a single payment platform. This integration is vital for the commuter to have a seamless trip planning and trip payment experience. It is also a strategic tool to draw more users into the system and adopt smart city services.

Integrated ticketing means the commuter can select and pay for a complete trip with multiple vendors and modes of transportation on a single app or website platform [5].

To provide this integration, a smart city manager must evaluate all the payment technologies to determine which method best fits the platform they are developing.

As mentioned in a previous section of the book, a challenge encountered when developing the energy or transportation system in a smart city is the aggregation of the numerous service suppliers.

We have shown how blockchain can address this issue by utilizing a decentralized ledger that can be viewed and updated securely. The fact that blockchain has a decentralized system does not mean that multiple vendors who may be competing will have to share their proprietary company information. This solution allows the service providers to cooperate on important services while maintaining individuality to provide a win-win solution for the customer.

2.9 Trends and Innovations in Payment Technology

Now that we have introduced the payment methods in general, and then discussed some payment technologies within the smart city context, we will look at some trends in the marketplace.

2.9.1 Contactless Payment [2]

Contactless payment systems have become popular in countries like the UK and Australia. Contactless payment involves using a chip-enabled credit or debit card or smartphone app to make payment at a contactless enabled terminal device by utilizing radio frequency identification or near-field communication (NFC) [4].

Contactless payment does not require the user to swipe the card or enter a pin.

Some types of contactless payment are:

1. Apple Pay
2. Google Pay
3. Fitbit Pay

2.9.2 Apple Pay

Apple Pay is an Apple iPhone app that allows the user to capture their credit card information onto the app and store it for use later with the iPhone for contactless payment.

2.9.3 Google Pay

Google Pay is an app for users of Android smartphones that is used to capture and store credit or debit card information on the smartphone for use to make payments at retail locations or transit systems.

The camera on the smartphone is used to obtain the card data by the app.

2.9.4 Fitbit Pay

As the name suggests, owners of a Fitbit wearable device may use the app to store their credit card information to use for contactless payment with Fitbit.

2.9.5 Smart City Payments

A smart city is a city that employs information and communication technology (ICT) to solve urban problems and better the city. The government strives to give essential services to the citizens and cater to their welfare in a smart city.

Different authorities have defined what a smart city is or what makes a city “smart” to them. IBM, the International Business Machines, and others have developed their own definition of the smart city. The consensus is that a smart city utilizes information and applies technology to address problems and resolve problems of the urban regions.

The vision of a smart city is to improve the functionality of the city. In addition, the government of a smart city should endeavor to enhance economic growth and boost the standard of living of the citizens through smart technologies and data analysis. The value of a smart city depends on how it utilizes the technology at its disposal rather than the availability of the technology.

To determine if a city is a smart one, you must look for certain features. These features include:

- Technology-based infrastructure.
- Progressive city plan.
- People living and working with available resources.
- Great initiatives within the environment.
- Functional public transport system.

For a smart city to thrive, public and private sectors with the environment must maintain good and healthy relationships. The two sectors must work hand in hand to create and sustain a data-driven environment.

For instance, the government can help install smart surveillance cameras in the environment, but the effectiveness of the cameras requires some technological contributions from many tech companies. So, both sectors (public and private) must commit so much work to ensure a functional smart city.

A smart city holds technology in high esteem. But apart from using technology, it needs many data analysts to assess and monitor the data in the smart city system. With this, addressing problems and offering suitable solutions will be easy and will ascertain the better living of its citizens.

2.10 Electronic Payments

Smart cities are often associated with cashless payment systems. However, there is no direct connection between them. Cash-based smart cities tend to perform better in terms of smart city rankings, while those that encourage cashless payment see fewer benefits and more challenges [6].

The KPIs included in the standard focus on debt service ratios; a city's financial health is determined by many factors, such as its ability to attract investors and businesses. The ITU has developed standards for smart cities, including establishing a system of electronic payments that has made it easier for residents to pay for products and services digitally. Financial companies are also part of the development process in ICT-based cities.

The ISO (International Organization for Standardization) has identified the importance of financial performance in smart cities, but the ITU has focused specifically on financial services.

2.11 Why Are Digital Payments Important in a Smart City?

According to famed British zoologist Desmond Morris, cities are nothing more than human zoos. Today, nearly 55% of the world's population lives in urban areas, and this number is projected to reach 68% by the year 2050.

As the population of urban areas continues to grow, there will be an increase in respiratory diseases from pollution. There will also be greater demand for new infrastructures to be built and upgraded—roads, power stations, and more. To keep urban centers comfortable and livable, further development in environmental quality, financial development, and infrastructural developments is a must. One way to do this is by developing a smart city.

Smart cities use cutting-edge technologies to optimize city functions, creating a better quality of life for citizens and increasing economic growth. These technologies include artificial intelligence, the Internet of Things, digital payments, communication technology, and information technology.

2.11.1 Role of Payments in a Smart City

Internet of Things payments are just as important for smart cities as other sectors. More and more governments and municipalities are realizing that to achieve the vision of a smart city, they must upgrade their current payment technology.

Digital payments can be essential in improving the entire ecosystem of payments. The use cases cover multiple C2G (citizen to government) transactions, such as electricity, tolls, water, taxi, education, tourist spots, healthcare, etc. Digital payments have the potential to bring more transparency to transactions and allow users to track their expenses. It is a cost-effective solution with instant payments that is sustainable and scalable.

The system can also be used for multiple governments to citizen (G2C) payments, including monetary awards, student scholarship, and subsidies. These use cases lengthen the system's reach and ensure better security than the current cash-based methods.

However, it is not without its share of internal and external challenges that make large-scale adoption quite difficult. Here is a magnified view of these challenges.

2.12 Digital Payments in Smart Cities

There are several challenges and opportunities associated with smart cities that a practitioner should be aware of. Some of these are introduced here and will be discussed in more detail in Chap. 4 of this book. Skowron [1] addresses some challenges with payment in smart cities.

1. *Internal challenges*

Many developing nations across the globe have not caught up with technological advancements and, as a result, are missing out on the benefits of digital payment options. It may be necessary for government bodies to rethink their methods for collecting and processing payments to offer consumers a more secure digital payment experience.

Most companies face the problem of constantly having to incorporate new tools and services into their systems while at the same time remaining flexible enough to meet ever-changing consumer demands. The lack of information sharing among departments, as well as the lack of a unified strategic vision for payments digitization, can make the process less effective.

2. *External challenges*

Poor connectivity, low Internet penetration, and poor infrastructure in developing countries adversely affect customer experience for online transactions. Many consumers are hesitant to engage in online transactions because of a lack of awareness about the safety and security of the payment system. Consumers with special needs encounter problems when making purchases online because they do not receive adequate training from online retailers, who are sometimes unable to provide support when shoppers need it most.

Due to the absence of a digital payment promotion campaign, people are unaware of how digital payments work. That causes them to feel hesitant about using digital payments and reduces the rate at which people adopt them.

Within the context of a smart city, some of the payment types mentioned above have advantages due to speed of processing, access, and general convenience.

With the proliferation of cellular phones and mobile payment apps for smartphones, mobile payments appear to be well positioned as the preferred payment method for a smart city.

Although many people think about mobile payments in terms of just a single method of payment, there are several ways of payment processing on a mobile application. One method is mobile web payment, also called WAP. Another is through SMS transactions, and a third is through contactless or near-field communications.

New innovative credit card payment technologies are emerging, with Mastercard developing a "virtual" card which they claim to provide instant payments for merchants. This type of card also moves from a physical card to a digital card which can be accessed from any mobile device.

For the smart city to be successful, the administrators must be willing and ready to form public-private partnerships with these technology companies since they have the expertise in developing technical products. At the same time, the technology companies must join hands with the city agency administrators to benefit from the market opportunity of the smart city.

In addition to the direct benefits of providing this technology to the governmental administration of the smart city, there is also the added benefit of the large amount of transaction data that mobile payments and other financial technologies generate. This data can be used for marketing purposes and presenting the customer with related products.

In entering these public-private partnerships, some consideration must be given to the privacy of consumer data, as there are legal boundaries which must be adhered to for the protection of the citizens. A primary concern is determining who owns the data and how it is used.

References

1. Skowron, J., Flynn, M. *The challenge of paying for smart cities projects.* Deloitte. Retrieved from the world wide web. https://www2.deloitte.com/content/dam/Deloitte/global/Documents/Public-Sector/gx-ps-the-challenge-of-paying-for-smart-cities-projects1.pdf
2. Aveni, T., Joep Roest China's Alipay, & WeChat Pay. (2017). *Reaching rural users*. CGAP World Bank Document.
3. Belgavi, V. (2017). *Creating Smarter Cities through digital payments*. Payments Newsletter. Retrieved from the world wide web. https://www.pwc.in/assets/pdfs/consulting/financial-services/fintech/point-of-view/pov-downloads/payments-newsletter-creating-smarter-cities-through-digital-payments.pdf
4. Trütsch, T. (2020). The impact of contactless payment on cash usage at an early stage of diffusion. *Swiss Journal of Economics and Statistics, 156*(5), 1–35. https://doi.org/10.1186/s41937-020-00050-0
5. Szabo, N. (1994). *Smart contracts*. Nick Szabo's Papers and Concise Tutorials. https://www.fon.hum.uva.nl/rob/Courses/InformationInSpeech/CDROM/Literature/LOTwinterschool2006/szabo.best.vwh.net/smart.contracts.html
6. Nakamoto, S. (2008). *Bitcoin: A peer-to-peer electronic cash system*. https://git.dhimmel.com/bitcoinwhitepaper/v/a5f36b332cb6a5fa9e701886f376ac1ac2946d07/.

Chapter 3
Electronic Toll Collections and Smart City Payments

3.1 Introduction

As mentioned earlier in the book, this chapter will address the technological aspects of payment systems. Here we will investigate electronic toll collection in more detail and electronic ticketing for transit payments.

Smartphones have become a ubiquitous part of the cultural landscape in the USA and across the globe. With increased capabilities and advanced electronic circuitry, the smartphone has been used as a camera, video recorder, and web applications service provider. There are "apps" for just about every computer function available, including banking transactions and e-commerce.

Not surprisingly then, applications have been developed by transportation agencies to allow commuters to pay for trips on toll roads and managed lanes. Several states including the state of Georgia have developed and used smartphone applications to allow commuters to pay for their travel on Interstate 85 through metro Atlanta.

On Interstate 85 in Georgia, the managed lanes are subdivided into component parts, where users can embark and disembark at various entry and exit points. These points are clearly marked along the highway, and the commuter is only charged for the section of the managed lane which they traverse. The payment system can calculate this congestion price and charge the commuter accordingly. This contrasts with other managed lane systems, such as Interstate 95 (Phase 1) in Miami, Florida, where the commuters are charged one fee for the entire distance of the high occupancy toll (HOT) lane.

Other managed lane systems have also developed smartphone apps for payments and transaction processing.

For instance, Florida HOT lanes and turnpike use SunPass for which it has developed an application. In North Carolina, the highway system uses Quick Pass for payment processing. Georgia residents who use Peach Pass can use the

© The Author(s), under exclusive license to Springer Nature Switzerland AG 2023
D. Graham, *Integrated Electronic Payment Technologies for Smart Cities*,
https://doi.org/10.1007/978-3-031-38222-2_3

smartphone app in both Florida and North Carolina which allows them the convenience of not buying a new tag when traveling to these neighboring states. Interoperability has been a key initiative of the Department of Transportation, with several states collaborating to use the same protocols for their devices in an effort to integrate the payment systems across state lines and for the ease of use and convenience for the commuter.

The smartphone apps such as Peach Pass mobile app allows the user to [1]:

- Set up a new account
- Add vehicles to each account
- View recent transactions
- View balances
- Access statements

The app does *not* allow for the removal of vehicles from an account. Any vehicle using the managed lanes is required to have a Peach Pass account even if they fall into the category of "exempt." These "exempt" vehicles include city transit vehicles, alternative fuel and electric vehicles, as well as emergency vehicles and motorcycles and carpools with three or more occupants (source: www.peachpass.com) (Fig. 3.1).

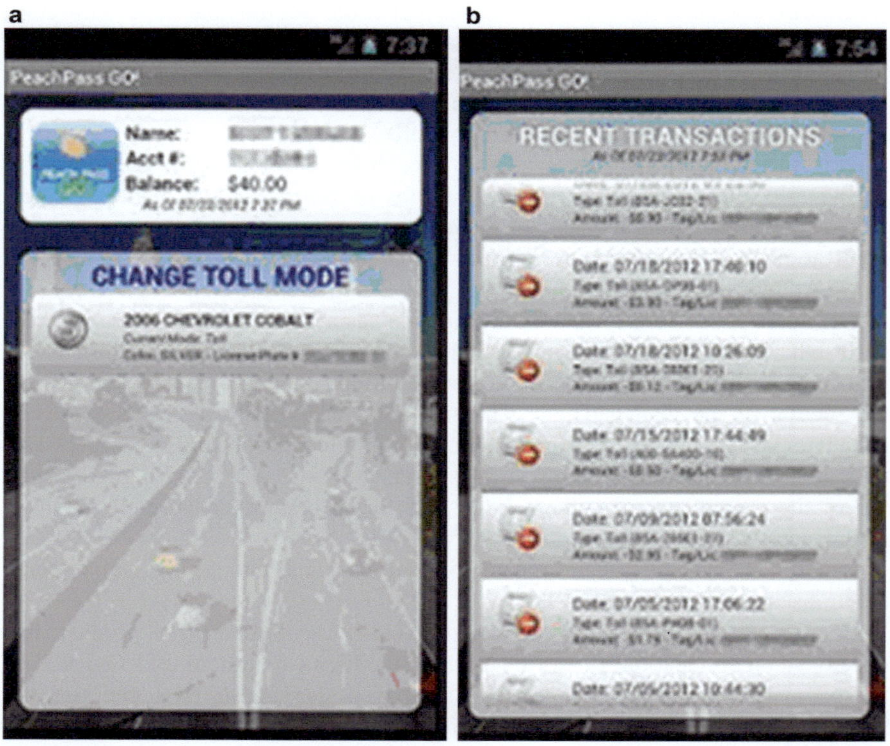

Fig. 3.1 Peach Pass mobile interface (**a**, **b**) www.peachpass.com

3.2 Electronic Toll Collection

Electronic toll collection (ETC) is a method of payment of road charges used on toll roads with a RFID (radiofrequency ID) transponder in the USA and internationally. With electronic toll payment, vehicles use a RFID transponder which is typically affixed to the front windshield of the vehicle. The account information and balance are encoded in the RFID tag (transponder) and can be communicated to a reader when the vehicle passes within the proximity of the antenna of the reader which is set up along the highway over the payment lanes.

The reader generates an electromagnetic field which is capable of interrogating and reading RFID tags on the passing vehicles at a high rate of speed.

The reader-antenna combination can also read multiple tags within its range at the same time.

Once the vehicle passes the reader, the information on the tag is processed by the reader's computer, and the toll is taken from the tag holder's account. The transaction is also recorded on the RFID tag and the information is displayed by indicators which show one of three states:

1. Toll paid
2. Toll not paid violation
3. Balance low

The advantage of using electronic toll collection (transponder) is that drivers do not have to stop at a toll booth to pay the fee but can continue through the toll gantry at a high rate of speed. For congestion pricing facilities, this is critical. Toll plazas require vehicles to stop, form a queue, and get service to pay the toll. All these steps require an increase in travel time and increase in congestion which counteracts the purpose of all congestion pricing facilities.

Thus, electronic toll collection with a transponder solves this congestion problem by allowing free-flow speeds through the HOT or Express lanes while processing the transaction and collecting the fee for the convenience of reduced travel time and increased speeds compared to the general purpose, non-tolled (free) lanes.

Electronic toll collection is used on such high occupancy toll (HOT) lanes as I-95 Express, in Miami, Florida (SunPass), and all throughout New England (NY, NJ, MA, PA) with E-ZPass transponder systems.

Electronic toll collection eliminates the need for a brick-and-mortar structure to house toll booths. It also does not require the need for toll attendants, thus reducing the cost related to those factors in the overall fee structure.

Radiofrequency identification (RFID) systems are also used widely for parking lot payments and access to restricted areas such as facilities for security personnel and high value storage.

Electronic tolling facilities report a 99.9% accuracy rate for reading tags and processing transactions correctly.

Many electronic tolling facilities in the USA currently use a different frequency band for their readers and tags, thus also a different protocol. However, work is being

Fig. 3.2 SunPass toll plaza with ETC and cash lane

done to improve the interoperability of FastTrack, E-ZPass, and SunPass ETC systems. This would allow for one protocol and drivers in California to use the same RFID tag when traveling cross-country to the east coast and to the NY, NJ, and MA vicinity.

The seamless integration of multiple ETC facilities would be a big boost to the transportation industry and help improve the congestion problem even more, since out of town users can more likely take advantage of the electronic toll facilities when traveling to new locations.

Electronic tolling facilities also utilize digital cameras in conjunction with computers to identify violators and issue fines. These systems implement automatic plate number recognition discussed in another section.

In this way, the system can both process toll transactions and identify violators using the computer database.

In fixed price toll facilities such as a turnpike, where all vehicles are tolled, drivers who have an electronic RFID tag with an associated account will pay less than the other drivers who don't have a transponder tag or account (Fig. 3.2).

3.3 Electronic Payment for Mass Transit

In previous sections of the chapter, we discussed electronic toll collection for travel on highway or toll roads. Another mode of urban transportation is mass transit, which includes light rail, subways, and buses. The payment technology which is

Fig. 3.3 Mass transit electronic card with magnetic stripe (NYC)

commonly used for these systems is that of electronic cards which possess a magnetic stripe on the back with the cardholder's account information and balance of funds stored. When the user swipes the card through a reader at the entrance to a subway station, the fare is deducted from the card, and the user gains entrance to the train. Similarly, for bus transportation, the user may swipe the card through a reader on the bus to gain access to the bus service after the fare is deducted from the card. Each swipe of the card deducts the fare from the balance stored on the magnetic stripe until the funds run out and the card is no longer valid (Fig. 3.3).

3.4 Pay as You Go Tolling

Another technology for electronic toll collection is the pay as you go device. This device allows commuters to install or attach a device in their car to record their miles driven, location, and other global positioning system (GPS) information. At the end of the month, any miles driven along toll roads are used to calculate the toll (congestion price) for those managed lanes, and the driver is sent an invoice which can be paid electronically.

The State of Oregon has initiated a program to allow drivers to participate in this method of electronic toll collection. The device (also called a dongle) is plugged into the vehicle's OBD (on-board diagnostic) system port and records data as the car is driven. At the end of the month, the driver then removes the device from the vehicle and connects it to a reader which reads the data and calculates the toll payment and generates an invoice (see www.roadchargeoregon.org).

This plug-in device (dongle) has also been used by some insurance companies as a means of monitoring driver behavior and using this data to reduce insurance premiums for "good" drivers.

In the Oregon pilot program, drivers are charged a per mile rate of 1.7 cents/mile instead of the 34 cents per gallon gas tax. This method of tolling is seen as a fair and equitable approach since users only pay for what they use and everyone bears their share of the costs associated with travel on the roadway. Drivers can use several possible devices for the collection of data and for transaction processing. Several of the devices utilize GPS which report location data along with miles traveled [2].

Since location data can be seen as intrusive by some commuters, at least one device is available to track the miles driven without revealing the location data.

The Oregon program of pay per use tolling has been used internationally. Other states such as California have initiated legislation to repeal the gas tax.

In the pilot program, 5000 drivers were initially enrolled in Oregon to install a device in their vehicles which would track how many miles they drove each month.

The device which they installed ranged from a simple insurance-type mileage tracker which is connected to the on-board diagnostic (OBDIC) computer port installed on all newer vehicles up to GPS-enabled smartphones which can record not only how many miles traveled but also the location of the vehicle during each period.

The pilot program developed originally in 2012 has been enacted into law, and in 2013, the Oregon senate passed SB 810 which allows for continued legal use of the road usage charging program.

The road usage charge will apply only to owners of vehicles which have mileage ratings of 20 mpg and higher and only to late model vehicles. The increase of electric and hybrid vehicles which attain these higher mileage ratings has significantly reduced the revenue from the gas tax in Oregon, and the road usage charge is seen as a way to replace some of this lost revenue.

Miles recorded by the device are reported on an open system, which is managed by a third-party vendor.

According to the Oregon Department of Transportation (see www.roadchargeoregon.org), there are three options for the technology used to record and report the miles driven.

- Basic. This system (described above) uses a simple device plugged into the vehicle's computer port and records only miles driven.
- Advanced. This system uses a GPS device which records miles driven and location. This allows for determination of whether the miles driven were within the state of Oregon or not. Miles driven outside the state of Oregon or on private roads inside of Oregon are not eligible to be charged under the road charging program.
- Smartphone. This system also uses GPS to log total miles driven much like the basic device. It also uses the GPS capability of a smartphone to determine which miles were driven outside of the state of Oregon and does not charge for these miles.

However, it does not distinguish (yet) between miles driven on private or public roads inside the state.

3.5 Toll by Plate

Toll-by-plate technology is an integrated payment system which uses high-definition camera technology and advanced algorithms to charge drivers for their usage of the toll road. In this system, the driver passes an array of cameras which can recognize a license plate on a vehicle. The image captured by the cameras can also be analyzed to decipher the alpha-numeric symbols on the license plate to "read" the tag. Once the tag is read, the fee for usage of the toll road or managed lane can then be assessed, and a bill is sent to the owner of the vehicle assigned through registration records [2].

Due to recent advances in artificial intelligence, the system can use pattern recognition algorithms to determine the likelihood of a shape representing any alpha-numeric symbol. Once this pattern is "recognized" individually for each shape, the values are combined to determine the complete license plate number.

Toll-by-plate systems have been widely deployed across the USA and Canada as an effective means of integrated payment systems technology. Since drivers do not have to stop the vehicle to have their plate read, it is also very convenient for the driver and the flow of traffic on a highway system. This method prevents the congestion which is often seen around toll plazas on such facilities such as a turnpike or a managed lane facility.

An added convenience of toll-by-plate technology is that the license plate can be read day or night in good weather or bad weather conditions.

Automatic number plate recognition technology (toll-by-plate) uses optical character recognition (OCR) which is used to recognize and record the license plate number for vehicles passing a certain point [3].

Optical character recognition has been used for years in the document processing industry for scanning and faxing documents. Optical character recognition is a technology which takes photographic images and transforms it into digital format. In using OCR for automatic number plate recognition, the system must be capable of reading images which are moving at high speed. Vehicles passing through a toll collection area using a RFID transponder will not stop or slow down and may be traveling (legally) at speeds of up to 75 mph. The OCR system will need to capture a clear image which is not blurred and record a license plate number.

OCR technology works with algorithms to analyze images or documents. In the case of a large document, the OCR algorithm will take a single page and break it down into blocks of sentences or lines. These lines are further broken down into individual words which are sets of characters grouped together. The words are then broken down into single characters which are passed through a matching procedure. These characters are then further broken down to a matrix of pixels like a television display. The OCR program will have a stored database of characters and associated images which it uses to match the external image. The system will then use the algorithm to predict what the probability is that this externally read character will be the same as the stored character or letter of the alphabet.

In the case of ANPR, OCR technology can be used to identify the characters and numbers on a license plate.

OCR technology has a wide range of applications. Some of these are:

1. Border control
2. Access control
3. Toll collection on highways
4. For law enforcement/parking enforcement purposes

In the case of law enforcement, automatic number plate recognition can be used to identify vehicles. A digital camera mounted on the law enforcement vehicle captures an image of the license plate as the vehicle passes by. If a match is found, the computer will immediately display the owner and other relevant information. Automatic number plate recognition can be used in combination with border patrol to protect land areas [2].

In the field of transportation, automatic number plate recognition is used to help collect toll on toll roads where vehicles are not required to stop at a toll booth and pay the toll.

Automatic number plate recognition systems can also record license plate information, at high speeds or in bumper-to-bumper traffic.

Because of these requirements, automatic number plate recognition systems are equipped with special cameras and lighting technology.

As an example, the distance moved by a vehicle traveling at 70 mph is great enough so that with a conventional camera the image will cause blurring and destroy the ability of ANPR to resolve the license plate number effectively.

In addition, license plates are generally coated with a reflective material to aid in the direction of light that is reflected by an illuminated source.

Automated number plate recognition systems utilize infrared light to aid in illumination and image capture at nighttime and in low-light conditions. The use of these technologies normalizes the detection and image capture quality so that the system will work properly in a variety of conditions which could adversely affect the performance of automatic number plate recognition (ANPR) systems.

Automatic number plate recognition systems use different algorithms to assist in the detection, processing, and output of captured images which help to resolve problems on license plates with unclear data.

Segmentation algorithms can be used to separate characters and groups of characters on a license plate.

After the boundaries between characters are determined, segmentation can be used to then enhance the different segments and remove undesirable sections of the image (such as dots and lines) which are not part of the license plate information.

3.6 ETC Communication Networks

The data collected by the electronic toll collection reader data is processed by a computer system located in a data center hosted by the transportation center at an off-road location.

This data must be transmitted through a communication network to the data center, the transaction (funds deducted from account) processed, and the new balance recorded on the customer's account.

The communication network provides the connecting link between the highway RFID reader and the computer data center and control center.

This communication network typically consists of a roadside unit in a cabinet on the highway. The equipment in the roadside unit will contain an electronic gateway which transmits the data using wireless protocol (IP) via a base tower (station) to the distant control central computer. The central computer (back office) performs the transaction (deducting funds from the account on file and updates the information on the card.

Later in this book we will formally introduce some network protocols which are used to transmit data wirelessly from the roadside unit to the control center through an electronic gateway. Some typical electronic protocols and gateway combinations include MQTT, Zigbee, Arduino, and CoAP.

The local transportation agency must make some decisions regarding which communication network and protocol is best for their location. Each reader and transmitter combination can have advantages and disadvantages which must be weighed carefully when designing the system. Some types of issues encountered include accuracy of the reader at night and in bad weather such as fog or snow. Since the transponder includes identifying information, privacy issues must also be considered when developing an electronic toll collection (ETC) system (Fig. 3.4).

3.7 Global Position Systems (GPS)

Global positioning system (GPS) is a technology which became widely utilized over the last 20–40 years. Having been used by the US military for many years, it was commercially introduced in the private sector in the 1990s.

Global positioning systems are used in several industries. GPS are used for vehicle fleet (such as rental cars) control and location tasks.

Global positioning systems (GPS) are used in dynamic pricing and integrated payment strategies to help reduce congestion in urban, metropolitan regions. GPS has been used in the trucking industry in Europe for several years to charge for use of toll roads and congestion pricing facilities. The use of GPS in place of a RFID transponder tag allows for electronic payment without the infrastructure of the RFID reader or antenna.

Fig. 3.4 Overhead toll gantry with RFID wireless system

Although the initial cost for GPS is higher than transponder-based systems (several hundred dollars for GPS compared to $20 or less for the RFID transponder), the overall system cost is typically lower. GPS is used by aircraft for precision guidance to destinations, thus reducing cost of fuel and saving time. GPS is used also for agricultural, mining, and earthquake monitoring systems.

Each GPS satellite circles the earth twice per day, providing constant coverage for signal transmission and reception [2].

Electronic toll collection is facilitated by GPS by having the user's location recorded as they traverse the highway system. The exact location of the vehicle is reported by the GPS, and when the vehicle enters a managed lane or toll road, the electronic toll collection system starts tracking the distance traveled on the toll road.

GPS is used to provide the precise distance and time usage on the highway, and based on this distance, the toll is assessed to the driver. As mentioned, since GPS facilitates this electronic tolling, no roadside equipment is necessary.

3.7.1 How GPS Works

Each global positioning system satellite broadcasts a radiofrequency signal with its exact location and time. This signal travels at the speed of light (c = 186,282 miles per second) to the earth where it can be detected by a GPS receiver. The receiver records the time the signal is received, and the information is sent by the satellite.

The GPS receiver uses this information and signal reception time to calculate its distance from the satellite which transmitted the signal. Once signals from at least four satellites have been received, the GPS receiver can use triangulation (a method from geometry) to determine its exact three-dimensional location.

Besides the location services, GPS also provides navigation service for subscribers to direct them from origin to destination.

3.8 Loop Detectors

Loop detectors are traffic detection systems placed under the surface of the roadway to collect information about the presence or passage of vehicles at certain points. Loop detectors not only detect presence but can measure occupancy, speed, density, the presence or absence of queues, and other physical characteristics of traffic flow.

Loop detectors can come in several different types depending on the technology which it uses.

According to the US Department of Transportation (FHA), some common types of loop detectors include [4]:

1. Inductive loop detector
2. Pressure loop detector
3. Magnetometer
4. Pushbutton
5. Magnetic
6. Radar
7. Sonic
8. Radiofrequency

Although there are loop detectors using the above technologies, not all these types of loop detectors are traditionally in use today, and only two of these are commonly used. These are the inductive loop detector and the magnetometer loop detector.

Inductive loop detectors work by installing one or two loops of electronic cable in the surface of the roadway where traffic needs to be detected. These cables are connected through a pull box at the side of the road to an amplifier and/or controller system.

The cables generate a magnetic field across the path of the vehicles crossing that point on the roadway.

When vehicles cross that point in the roadway, they disturb the magnetic field and modify its intensity.

This change in field intensity can be detected by the controller and indicates either the presence or passing of a vehicle over the loop detector's cables.

In signalized intersections, where multiple loop detectors are present, the magnetic field of one loop detector can sometimes interfere with the fields of another one and cause a phenomenon called "crosstalk."

When crosstalk occurs, a loop detector can possibly mistake the interference from another loop detector and interpret this as indicating the presence or passing of a vehicle.

In order to prevent this type of mistake from occurring, each loop detector in proximity of each other is set to operate at a different frequency so that the magnetic fields do not interfere with each other.

Loop detectors are used for a variety of applications to control traffic at intersections. Some of these applications such as priority vehicle intersection control can be used to allow emergency vehicles a free passage through an intersection when the signal is red. Thus, fire trucks and ambulances can modify the signal at the intersection using the access to the loop detector system, to change the lights so that it is not necessary for them to stop at the light.

Loop detectors can also be used to coordinate the signals at multiple intersections so that the traffic leaving one intersection with a green light will experience a green light at the time it arrives at the next intersection. This allows for the reduction in congestion in busy downtown city areas and improves the flow of traffic.

Loop detectors are also used on managed lanes to determine density and occupancy levels at specific points on the highway. The electronic payment system then uses these density figures to calculate the dynamic price for a time interval.

Loop detectors can also be used to assist with pedestrian crosswalks, by using the input of pedestrian presence; the loop detector can modify the signal to allow for pedestrians to cross busy intersection streets.

3.9 Dedicated Short-Wave Communication (DSWC)

Dedicated short-wave communication (DSWC), also called dedicated short-range communication, is a technology which has been developed to enable vehicles to "talk" to each other and to the electronic receivers on the road infrastructure.

Dedicated short-wave communication consists of a combination of GPS with radio transmission through a receiver and transmitter system. The US Department of Transportation (see www.transportation.gov) has allocated a 75 MHz frequency range on the 5.9GHz band for the operation of DSRC in the USA. In Europe and Japan where DSRC systems are in use, the frequency range is different; thus, the systems are not compatible with each other. DSRC will transmit information up to 300 meters to nearby vehicles or infrastructure units with receivers in the same frequency range.

Vehicle-to-vehicle (V-2-V) communication can be useful in a variety of transportation applications.

Some information which is transmitted from one vehicle to another include (see transportation.gov):

- Vehicle GPS location
- Vehicle speed

- Vehicle acceleration
- Vehicle direction of motion
- Vehicle transmission (gear) information
- Braking information

No personal information such as license plate number is transmitted between vehicles.

Vehicles can also transmit and receive information to electronic equipment placed at strategic locations on the roadway. This type of communication is often referred to as V-to-I, or vehicle to infrastructure communication.

Thus, by using V2I, the system can detect congestion or accidents along a roadway and broadcast this information to other vehicles to allow for intelligent rerouting of vehicles from the affected areas.

Vehicle to infrastructure systems is applied in the following areas:

- Emergency vehicle prioritization
- Ramp metering
- Intelligent transportation system (ITS) alerts
- Work zone warning
- Traffic congestion and weather information alert

These are just some of the applications which V2I can be used to improve the efficiency of transportation systems.

V2V systems provide several safety systems which can be used to prevent accidents by warning the driver electronically of impending danger. These warning systems, however, do not take over operation of the vehicle or control of the vehicle at any time but simply give the driver an increased chance of avoiding a collision.

Some of the V2V electronic warning systems include:

- EBL: emergency brake light warning system (allows driver to know if vehicle ahead has stopped suddenly or is stopping, without seeing the vehicle's brake light)
- BSW: blind spot warning (warns drivers of vehicles present in a blind spot)
- DNPW: do not pass warning (warns drivers of incoming vehicles which will cause a collision if the driver passes a slower-moving vehicle which is in front of the driver).

Intersection warning provides a warning signal if a fast-moving vehicle is about to cross the driver's path in an intersection and can cause an accident.

3.10 Autonomous Vehicles

Autonomous vehicle technology (self-driving cars) is indirectly related to payment technology and congestion pricing since it can form platoons moving at high speeds in a HOT lane. These platoons can increase the capacity of the roadway or lane being utilized. The concept of a car that can drive itself has been around for several years. The invention of cruise control which allowed for automatic maintenance of a preset speed brought the concept that much closer to reality.

In parallel with research into self-driving cars, airplanes have developed and implemented autopilot for commercial airlines. On long transcontinental flights, the autopilot can be used to ease pilot fatigue and allow for rest. This much needed rest can be useful for more arduous tasks such as take-off and landing.

Cruise control was invented in the 1940s and first used in GM cars in the late 1950s. With continuing advances in technology, improved systems such as adaptive cruise control have been introduced in later model automobiles. Vehicles with adaptive cruise control can slow down and speed up automatically to maintain a safe travel distance to a vehicle ahead of it.

This cruise control mechanism is a central part of the self-driving car. Self-driving cars can not only control speed but also control direction through steering mechanisms.

The steering of the vehicle to maintain a lane through a curve can be accomplished by several technological methods, but one which has been frequently used is that of a magnetometer and magnetic field (California PATH publication).

One such system (source: California PATH publication) is that of using magnetic markers in a roadway to provide positional reference of the road geometry and lateral and longitudinal position.

The magnetic field strength is detected by the magnetometer located inside the vehicle. The information about location and positional reference is processed by a special computer in the vehicle which can control the steering of the vehicle and adjust for changes in the magnetic field strength as the vehicle moves along its lane.

3.11 Case Study: Electronic Toll Collection in State Route 91 in California [4]

State Route 91 Express (also known as 91 Express) was one of the first congestion pricing facilities to open in the USA.

SR 91 Express is a 10-mile-long managed lane facility which operates between the county line of Orange/Riverside County and the I-55 (Fig. 3.5).

The facility opened in 1995 with an initial cost of $135 million (see https://www.octa.net/) by the California Private Transportation Company (CPTC) but was later acquired by the Orange County Transportation Authority (OCTA) in 2003.

The 91 Express was one of four projects enabled by state legislation (AB 680) in 1989. The SR 91 was the first project developed by this legislation, while two others were not developed, and the SR 125 in San Diego was the next project developed by the AB680.

This legislation allowed the California Private Transportation Company (CPTC) to build a toll facility in the median of the State Route 91 freeway in Orange County.

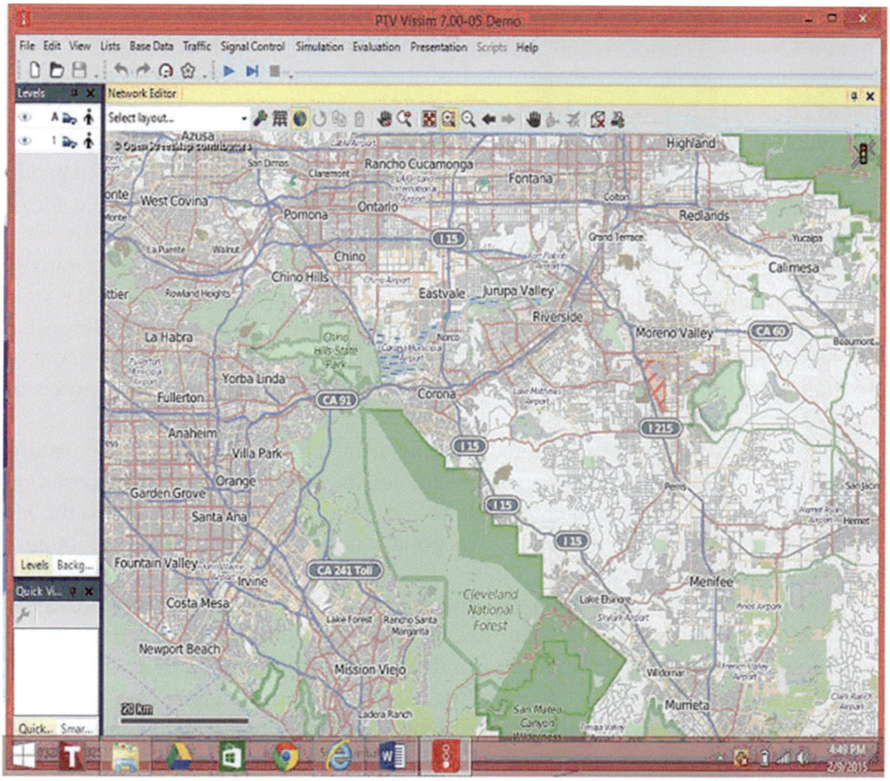

Fig. 3.5 Location of SR 91 in Orange County, CA (ptvgroup.com)

That legislation gave the CPTC the ability to enter into a franchise agreement with the state to build the tolled facility. That franchise agreement allowed the CPTC to build the SR 91 Express and charge tolls for a period of 35 years from the date of opening. Since they opened SR 91 in December 1995, they agreed upon a 35-year period until December 2030 to own and operate the facility. In addition, the private company CPTC had a noncompete clause which prohibited any improvement projects which could compete in the corridor. This inability to allow improvement projects led to a lot of frustration toward the middle and late 1990s as there was significant growth in travel demand projected for the next 20–30 years in the region and also significant growth occurring in the areas where the SR 91 Express was being developed. Even though the state could build an improvement project, they had to negotiate with the CPTC to build the improvement project. In one example, the Orange County Transportation Authority negotiated with CPTC to build a 3000-foot improvement project on the west end of the highway which, after analyzing the traffic and revenue studies, a figure of $4million was agreed upon (based on personal interview with OCTA representative June 2018).

With that extension experience and with the continuing growth in the region, the Orange County Transportation Authority (OCTA), along with Riverside County which is adjacent to Orange County (SR 91 Express connects Orange County to the Inland empire in Riverside County), joined together in planning for the increased demand. This increased demand occurred due to the relatively low house prices in Riverside County compared to Orange County. Because house prices in Riverside County are close to half of the home prices in Orange County, there are a lot of commuters who work in Orange County but live in Riverside County.

Another reason for the move to buy the Express lane from the private company (CPTC) is the fact that this company is not based in Orange County, or in the state of California. Instead, the headquarters for CPTC was based in Omaha, Nebraska. So, the process in which the proceeds from the toll revenue were being taken out of the state and transferred to another part of the country was a second source of frustration.

Due to the mounting frustrations resulting from high cost for road expansions combined with increasing transportation demand, the Orange County Transportation Authority (OCTA) decided to enter legal proceedings to buy out the right to the SR 91 Expressway. After about 2 years of negotiation, in 2003, the SR 91 was bought from the CPTC by OCTA.

After negotiations, the OCTA bought the SR 91 for a price of approximately $207 million (compared to $135 million which CPTC paid for construction) [4].

The purchase was valued based on a discounted cash flow basis calculated on projected revenue stream and cost of capital. Based upon the analysis, the eastbound direction (from Orange County to the Inland Empire) was priced higher than the westbound direction; this was because there were more people traveling to go home in the evening versus people traveling to come to Orange County in the morning.

The configuration is such that there are more lanes in Orange County and fewer lanes in Riverside County, and because of this funneling in the eastbound direction, congestion increases. On the other hand, because one moves from a location with fewer lanes to a location with more lanes, there is less congestion in the westbound direction.

After the January 2003 purchase, the OCTA removed the noncompete clause from the SR 91 and adopted a new toll policy. The new toll policy was derived from the previous toll policy and increased the prices depending on the congestion level on the Express lanes. The toll algorithm developed was based on an analysis from the traffic and revenue department which determined the volume of traffic which the Express lanes could accommodate, without a breakdown in flow. Given that one of the goals was for a free-flow speed around 50–60 mph, the optimum number of vehicles which could be operated safely without achieving "jam density" would be used to calibrate the congestion pricing scheme.

It is important to note that unlike the I-15 and SR110 where pricing is updated every 15 minutes (or less), the pricing on the State Route 91 Express is not considered a dynamic pricing scheme since it is not updated on an hourly or intra-day basis. Instead, the traffic volume and demand are reviewed every 3 months. At that review, the price is either increased, decreased, or stays the same based on the measured volume at that time.

The toll policy is based upon an algorithm which determines how much the toll is increased or decreased when reviewed. The toll policy will raise the price either $0.75 or $1.00 when reviewed, depending on the traffic volume. Similarly, the prices may be decreased by $0.50 if an appropriate decrease in traffic volume is attained.

Toll prices and volume peaked during 2007–2008 on the SR 91 Express. During the recession years after 2008, the volume decreased, by approximately 25%, and to combat that decrease in ridership, the tolls were decreased to attract more commuters to use the road. Over the last 3–4 years, the toll has not decreased due to the economy stabilizing.

The expressway contract is planned to be extended for another 30 years, and the Orange County together with Riverside County Transportation Commission have decided to extend the current SR 91 Expressway by approximately 10 miles east into Riverside County.

The current SR 91 Expressway runs from SR 55 (Costa Mesa Freeway) to the Riverside County/Orange County line, which is approximately 10 miles. The planned extension will add another 9 miles to the I-15 in Riverside County. The extension project which will be funded by RCTC will be similar in geometry to the current stretch of the SR 91, in that it will have two express (priced) lanes and be separated by plastic pylons from the general-purpose lanes. Although the major general-purpose lane of the extension is expected to be funded by sales tax, the express lane portion of the project will be funded by the toll revenue.

Since the extension of the toll road will include both Orange County and Riverside County, both counties have worked together on the legislation to approve the extension to SR 91 and to get the 30-year contract extension from the state. As a result, both the Orange County and the Riverside County portions of the SR 91 Expressway will have the same pricing structure and be distance-based as well as volume-based.

Most of the congestion occurs on the eastbound traffic during peak hours (around 3–6 pm) on weekdays. One of the goals of the expansion project is to correct the "funneling" effect by increasing the number of lanes in the eastbound direction in Riverside County, to match the number of lanes in Orange County.

In the current configuration of the SR 91 Express, there are two express lanes and five general-purpose lanes in the corridor. On average, there are approximately 285,000 daily trips on the SR 91 corridor of which the Express lane comprises about 12% of the total daily demand (~34,000) across all time periods. During peak period hour, the percentage of Express lane traveler volume to total volume is approximately 30%.

The primary source of funding for the SR 91 Express lanes is through the municipal bonds issued on the stock market.

The sole source of revenue for the project is the tolls or user fee on the Express lane. There are no tax dollars from state, federal, or other government sources used to extend the Express lanes.

In California, toll roads such as SR 91 use a transponder which is Title 21 compliant. Title 21 is a legal oversight on what components the transponder must have. The transponder used on the SR 91 is compatible with other toll facilities throughout California and can be used on the Golden Gate Bridge, the Bay area toll authority bridges, also in San Diego on the SR 125 or SR115, or in Los Angeles on the SR110. As vehicles pass under the RFID reader on SR 91, the system reads the transponder information to perform the transaction and pay the toll. If the vehicle does not have a transponder (or if the transponder is out of funds), the system will take a digital photograph of the license plate and the owner is issued a toll violation notice in the mail.

The system is equipped with an enforcement camera which is very precise. It can capture the correct image of the license plate under all weather conditions and at night in the dark. It is rated at a 99% accuracy level. The only issues which have been observed with the camera enforcement accuracy are that of vehicles traveling at high rate of speeds. For instance, at speeds over 100mph, there can be increased inaccuracy due to blurring which leads to bad license plate images. Another issue which can sometimes create a problem for the accuracy is license plates with a transparent plastic glass over the metal plate. This can sometimes produce a high glare which causes problems with the camera. In those cases where a bad image was captured by the enforcement camera, the system will discard this image and use other methods. Overall, the percentage of bad images to total images captured is extremely small.

The pricing scheme on SR 91 starts from $1.45 (minimum) and goes up from there. Typically, this toll is charged during midnight hours when there are the fewest vehicles on the road (lowest demand). The maximum toll occurs in the eastbound direction on Fridays from 3 pm to 4 pm and costs $9.85. During this peak period, the demand is known to have the highest average volume. Due to the difference in volume on the westbound traffic and the demand being lower compared to the eastbound traffic, the maximum toll on the westbound Express lanes is approximately $4.85. On the westbound lanes, the traffic operates at 60–70% of capacity, while the eastbound is at 90–100% of capacity.

The prices on SR 91 also vary by day. Thus, the price on Friday 3–4 pm is different from Monday 3–4 pm, since the expected volume is less on Mondays than on Fridays at the same hour.

Since the goal of the congestion pricing scheme is to reduce congestion during the peak hours, toll prices are less expensive just before and just after the peak hours. This will encourage drivers to travel during those "shoulder hours" instead of during the peak morning and afternoon hours.

In reviewing the weekly demand profile for the SR 91 Express, administrators find that Tuesday has the lowest peak hour demand and Friday will have the highest. Thus, Tuesday peak hours will have the lowest toll price and Friday will have the highest.

As mentioned previously, at the end of each quarter, analysis is performed to determine how the traffic volume has performed on an hourly basis. Thus, each hour of a particular day of the week is compared with the same hour and day for the previous 12-week period to determine if there is a consistent increase or decrease in volume. If the analysis shows consistent 75–100% capacity utilization, then an increase is recommended for that hour.

When a price increase is recommended, it will have a minimum change of $0.75. The maximum increase is $1.00. On the other hand, if a decrease in toll price is recommended, there is one amount of ($0.50) by which the toll price can decrease.

3.11.1 SR 91 Tolling Algorithm

The tolling algorithm which determines the price increase or decrease is based on the volume to capacity ratio. The Orange County Transport Authority (https://octa.net/pdf/SR-91Plan2021.pdf) has determined that traffic capacity should not exceed 3400 vehicles per hour on the two Express lanes on SR 91.

When the quarterly review is performed, each time that the volume exceeds 90% of the capacity (3128vph) for an hour, that event is recorded and added to a count. If the threshold of 90% capacity is achieved more than six times during the 12-week period, the average of all those numbers were calculated. If that average falls between 3200vph and 3300vph, the toll is raised by $0.75. If the average falls between 3300 and 3400 vehicles per hour, the toll is raised by $1.00 for the next 12-week period.

I-95 Express HOT Lane (Figs. 3.6, 3.7 and 3.8)

Fig. 3.6 Dade County, Fl I-95 North HOT lane (peak hour)

Fig. 3.7 Dade county, FL I-95 South HOT lane

Fig. 3.8 I-95 Express lane entrance (Miami-Dade, FL)

3.12 Blockchain Technology

Blockchain is a relatively new technology which has garnered a lot of attention over the last few years. With the exponential increase in the speed and storage capacity of computers, the blockchain concept has become a reality.

Blockchain which was initially developed as a way to monitor and time stamp digital data [5] has been adapted for use as a way to conduct financial transactions. In addition to Bitcoin and other cryptocurrency, blockchain is revolutionizing the way financial transactions are carried out.

In the current banking system, all transactions are carried out at a centralized location, and information is recorded by the bank. With blockchain technology, the financial transaction is carried out in a distributed network and all members of the network have the same information which must agree for the transactions to be valid.

Blockchain uses hashtags to identify each block in the chain. Identification information about the preceding block in the chain is also stored in the current block.

Once a valid transaction takes place in any block in the chain, all other blocks are notified and updated with the new data about the financial transaction and current status of the account.

With this new reliable way of conducting and verifying transactions over a computing network, many industries have adopted block chain technology as the way to conduct business transactions. Transportation networks including managed lanes have begun to examine crypto currency as a means of accepting payment for toll lanes.

It is not surprising to conceptualize a payment system in which drivers are allowed to pay the toll through an account with crypto currency. Such a payment system would utilize blockchain technology for digital financial transaction processing and verification.

References

1. Peach Pass. (2018, November). *Retrieved from the world-wide web*. www.peachpass.com
2. Graham, D. P. (2013). *A comparative evaluation of FDSA, GA, and SA non-linear programming algorithms and development of system-optimal methodology for dynamic pricing on I-95 express*. Doctoral Dissertation. www.etd.fcla.ucf.edu
3. How GPS Works. (2018, November). *Retrieved from world wide web*. www.transportation.gov
4. Sullivan, E. (2000, December). Continuation study to evaluate the impacts of the SR 91 value-priced express lanes. In *Final report*. Cal Poly State University, Department of Civil and Environmental Engineering, Applied Research and Development Facility.
5. IBM. *What is blockchain technology?* https://www.ibm.com/topics/blockchain

Chapter 4
Smart City Implementation: Challenges and Opportunities

4.1 Important Cases in Smart Cities

Many corporations are trying to create a sustainable cashless ecosystem, but few companies have done so successfully. The companies that have succeeded in this realm offer an array of mobile payment options and integrate different forms of technology into their platforms, two important factors that can help a company find success in the cashless space.

It is possible to completely overcome these challenges by boosting the adoption of digital payments. The following use cases provide a few examples of how these challenges can be overcome.

- *Citizen Facilitation Center*

Corporations are taking firm steps toward digitizing their transaction and payment processes. The days of visiting the brokerage or ward office to pay taxes are gone. Corporations have begun accepting digital payments, such as AEPS—a cashless payment system that uses social security data—and issuing mobile apps and POS terminals for tax payments. Reuter [1] reviews some human aspects of smart city implementation.

- *Direct Benefits Transfer Based on Aadhaar*

India's Aadhaar is not only an identification number, but it can also be used to transfer benefits to citizens; many municipal corporations in India have begun using Aadhaar-based prepaid cards to offer their residents essential services such as access to government subsidies or the ability to buy food at certain stores.

These Aadhaar-loaded prepaid cards can help students to buy school supplies and uniforms, further ensuring that their funds are only used for the purposes that have been permitted by the relevant corporation.

© The Author(s), under exclusive license to Springer Nature Switzerland AG 2023

D. Graham, *Integrated Electronic Payment Technologies for Smart Cities*,
https://doi.org/10.1007/978-3-031-38222-2_4

- *Doorstep Solutions*

A few governments have begun delivering their tax collection services from door to door, where an officer collects taxes from digital devices at a citizen's house on the citizen's chosen date and time.

The One City One Card System allows citizens to use a prepaid card to cover the cost of multiple services offered by their local government, including public transportation and necessities like water and power.

- *Increasing Online Payment Instruments*

With the rise of new modern payment methods like mobile wallets, UPI, EBPP, and IMPS, corporations are looking to improve their customer service experience. These methods allow customers to conduct transactions using their smartphones or personal computers [2] and can serve as a gateway to lower shipping costs and boost digital payments.

Multiple departments of corporations are also being moved to online platforms while integrating these departments with payment gateways for cashless tax collection.

4.2 Emerging Digital Payment Solutions for Smart Cities

We have only discussed a few of the ways that digital payments could be leveraged in smart cities. There are many others, some of which include:

- *Unified Payment Instrument*

Public transport like local trains, buses, and metros can be paid through this quite common digital payment mode.

It can also be integrated into a health card to allow trouble-free access to free medical benefits by citizens.

This common digital payment mode can also be deployed to benefit several segments of the society

- *Internet and Mobile Portals*

Markets and retail stores which are avenues for purchasing essential farming equipment and seeds can benefit from POS terminals as well as mobile devices that can enable transactions.

4.3 Grid Stability

The smart energy system must be stable. The energy grid must be able to consistently provide power.

In the new smart energy system, there has been a change in the hierarchy of the system. We no longer have a top down (vertical structure) where the power company generates the power in a single location then transmits the power over long distances to the city and where it is then distributed to the consumer. The transformed smart energy system is decentralized, and energy can be produced and injected into the grid at many points. This type of system results in a flat hierarchy where the end user can also produce power as well as use and buy power from the grid. The system is more transparent since the data provides information on where energy is being wasted and how this waste can be avoided. Law [3] addresses some of these challenges.

The system must also be synchronized because of all the new sources of energy, so that the grid is balanced and the right amount of power is placed on the grid based on the current demand, that is, the supply must match the demand even during peak hours.

Charging stations for electric vehicles and energy storage facilities present new benefits and challenges. With the recent buildup of new charging stations due to the influx of electric vehicles, governments need to catch up with the legal implications these facilities present. The stations must be regulated but allow for ease of usage, operation, and incentives for growth to facilitate expansion of electric vehicles and green energy. This is one way to reduce air pollution and fight climate change.

Integration of all stakeholders is necessary for a smooth operation of producing, consuming, and brokering energy resources; thus, a smart energy platform is needed to allow for the integrated functioning of the transformed system.

4.4 Grid Security

In 2021, there was a breach of security of a major pipeline in the USA. The colonial pipeline suffered a data breach by hackers who utilized ransomware to effectively shut down the pipeline for several days.

By using a single password, hackers were able to gain access to the colonial pipeline virtual private network (VPN). The security of the VPN did not use multifactor authentication which might have stopped the attack. The hackers requested and were paid several million dollars in cryptocurrency after threatening to shut down the pipeline. One of the lessons learned was just how easily a major energy producer can be shut down and result in severe chaos and disruption in the economy.

4.5 Data Pricing

A major consideration in implementing a smart energy grid is the cost. Who will pay for the cost to upgrade the grid? Should the city or should the utility?

Of major importance is the data that is generated by the smart energy grid. This data is valuable as it provides information on the customer's usage of electricity. This information can be used by energy brokers to tailor services to a specific community or individual. As such, the data can be sold to stakeholders and service providers. The pricing of this data is a point to consider when implementing a smart energy system.

4.6 Smart Transportation Systems

Smart transportation technologies in a smart city can be used to monitor and improve the flow of traffic on the roadways. LED (Light emitting diode) message boards on a highway, for example, are used to warn and inform drivers of road conditions and alternative routes. In case of an accident, the LED signage will advise the commuters of the issue and expected wait time. Alternative routes which can be used to bypass the accident scene may also be presented.

In the same way, several technologies have been developed which provide useful information to guide the driver from their origin to destination.

Some of these include:

1. Cameras
2. Sensors
3. Infrared devices
4. RFID
5. GPS

A second major area of application of transportation technologies is that of autonomous vehicles or driverless vehicles.

4.7 Autonomous Vehicles in the Smart City

The driverless vehicle has been in development for over 20 years. The advances (ICT) enabled the serious attempt to control a vehicle on the road in normal driving conditions.

Self-driving vehicles have been classified according to levels of automation:

Level 0: No type of driver assistance available
Level 1: 1 type of driver assistance available
Level 2: 2 types of driver assistance available
Level 3: Conditional autonomy (a combination of automated driving system and ICT)
Level 4: High autonomy (most driving completed by computer)
Level 5: Full autonomy

Driverless vehicles have some factors which must be considered within a smart city. The infrastructure of the smart city must have components for vehicle-to-vehicle communication (V2V). The driverless vehicle must also be able to communicate with the infrastructure (V2I).

The data protocols must be standardized so that vehicles from different manufacturers can communicate with each other.

The data transmitted from vehicles must be protected as hacking into the data in a driverless car could lead to disastrous results.

Questions related to responsibility for accidents will need to be addressed, and determinations about whether to hold the driver accountable or the car manufacturer must be answered.

4.8 Challenges and Opportunities

Implementing a smart city involves digitalization in several sectors such as energy, transportation, and logistics as well as financial systems.

In any new implementation for an industry sector, we will encounter both challenges and opportunities. Earlier, we introduced the energy sector and what considerations should be made when developing a smart energy system.

For each phase of the development, we find that there are challenges which must be overcome, and an assessment of the costs and benefits should be performed to guide the process. We want the smart city.

to reap the positive improvements from the transformation but not at an unreasonable long-term cost.

We will examine some challenges faced when embarking on a smart energy transformation project for a smart city and then look at the opportunities. Later, we will examine the transportation industry to determine what are the challenges and opportunities as well.

4.9 Challenges in Smart Energy

In a previous section of the chapter, we started to discuss some of the issues and considerations related to smart energy transformation. We discussed how the new digital economy has impacted the way energy is generated and transmitted to the customers.

Energy production has for a long time been done from a central location that electricity which is produced at the central location is then transmitted long distances (typically by overhead transmission) until it gets to the community in which it is to be distributed. Once the electric power reaches the city location, it is then stepped down from high voltages to a lower voltage in substations around the city.

The lower voltage power is then distributed to the community through power lines and transformers before connecting to the residential or commercial building.

In the new digital system, energy production has become decentralized with business and residence customers installing their own solar panels which feed into the grid. In addition to solar power, wind energy has become a major power generation source. Wind farms are dispersed across a wide area and often are located many states away from the city which the power serves. In contrast with an oil-producing site or nuclear power plant located in a single city, wind farms are decentralized and spread out over a wide range to maximize the weather-related wind generation.

Considering this change of decentralization, the challenge for the smart city is to aggregate all these sources and coordinate the different energy producers to supply energy to the smart city in the right amounts at the right times.

In many cases the challenges which come about from digitalization can often be best approached and solved by technology. Many electric utilities have installed smart meters which are capable of being read remotely and can detect the power consumption without the need for an in-person meter reading from the company. The smart meters can determine the net power usage if the location is generating their own power using solar panels and feeding that power back onto the grid.

4.10 Smart Energy Opportunities

The digital smart energy system brings several opportunities for the agency or utility as well. One of the immediate opportunities is the collaboration between the smart energy provider with other institutions to monitor and improve the sustainability and living conditions of the citizens in the smart city.

Another opportunity derived from digitizing the smart energy system is the data produced which can be sold or used to determine trends which can save the utility in reduced costs and thus increase profits.

4.11 Challenges in Smart Transportation [3]

Over the last 20 years, there has been a significant shift in the number and modes of transportation. With the expansion and application of ICT to transportation, there are the new concepts of smart transportation and smart mobility.

Major increases in the production and usage of electric vehicles, autonomous (driverless) vehicles, and car sharing have opened a new way of imagining urban transportation. These new modes of travel have been widely adopted by smart cities, and many populations have moved away from the concept of car ownership as the primary means of transportation in urban locations.

The state of California has passed laws requiring electric vehicles to be sold and used in the next 15–20 years to cut down on air pollution in the major cities. Major rental car companies have made huge investments in electric vehicles for the new fleet. These examples show that electric vehicles are not just a fad but are here to stay. In addition, the big three automakers (GM, Ford, Chrysler) have all started to produce their own electric vehicles or bought existing electric vehicle manufacturers to add to their portfolio of cars.

In addition to the emergence of electric vehicles (EV), there has been a large increase in companies offering charging stations for the EV. The automaker Tesla has had its own charging stations for several years and another one includes ChargePoint.

Even within the electric vehicle space alone, there is the challenge of how to standardize charging for all electric vehicles from the different manufacturers.

The smart city transportation network provides services using all three modes mentioned above but also includes mass transit buses, trains, and car hailing services such as Uber and Lyft. This can be a challenge when providing the commuter with a seamless trip from origin to destination. All these modes of transportation should be coordinated to allow for a smooth commute. This coordination will require some sharing of information on a single platform for the user to book their trips. The sharing of data can be a difficult idea for private companies who are competing for the same fare, and this hurdle must be overcome to make all the parts work together smoothly.

4.12 Integration

The existence of all these modes of transportation which can be booked digitally presents a challenge for the smart city planner. Echendu [2] studies some of these challenges faced in Sub-Saharan Africa. For the commuter to make the best choice for travel, the smart city should integrate all the services with the modes of travel and prices on a central platform. This can be done through an application for smartphone; however, it requires several independent entities working together and sharing information with competitors. One solution which is recommended in this book is the use of blockchain as a way of performing transactions on the integrated digital platform.

4.13 Regulatory Challenges

When autonomous vehicles emerged as a real possibility with companies like Tesla, Zoox, and others making major strides in technology over the last 10 years, serious questions arose about how to regulate these types of vehicles on public roads [3]. What are some standards the government should set for self-driving cars? Where can these cars be driven or parked? Several questions about accountability in case of an accident also must be answered in the case of fully self-driving (level 5) cars when being operated on the public roadway.

4.14 Smart Transport Opportunities

The implementation of technology in a smart transport system brings several benefits. One of these is the reduction in congestion. The large number of vehicles in a typical city results in travel delays daily. Commuting from home to work has become a source of time waste for most city dwellers [4].

Congestion not only causes travel delay but also results in serious levels of air pollution which leads to health problems or death.

The technology of autonomous vehicles and electric vehicles can make a significant impact on solving or reducing these problems.

Autonomous vehicles can utilize sensing technology to take up less space on the roadway.

This is called "training" where the cars are driven in a group near each other. Several studies have shown that a fleet of connected autonomous vehicles can speed up traffic by 30 percent or more [4]; this reduces travel time and congestion to the same degree.

Another opportunity of the smart city transportation system is the reduction in air pollution by using electric vehicles. In major cities, the gas-powered combustion engine generates significant amounts of air-polluting chemicals which not only sicken the residents but lead to global warming [1]. This opportunity to slow global warming is a driver for adoption of smart city technology in transportation.

The third opportunity in adopting smart transportation systems is it allows for less car ownership and more car sharing. With fewer people owning cars, there will be less air pollution and congestion on the city roads. Car sharing by using a digital platform such as Uber or Lyft allows for the residents of a smart city to complete all their travel requirements with fewer cars and less negative impacts to the environment.

The fourth opportunity for adopting smart transportation systems is the huge volume of data that is generated. The data from car-sharing apps, mass transit, and EV charging stations provides valuable information which the administration of the smart city can use for economic benefit.

The data can be used to learn travel demand trends and help in planning facilities and services for large groups of people.

4.15 Implications of Smart Transportation on Mobility

The new technology applied to vehicles and developed by car manufacturers into autonomous vehicles and electric vehicles has not only led to improvements in smart transportation but also fundamental changes to the way people think about mobility [3]. There has been a paradigm change where millions of people no longer see the need to own a car but find it more feasible to use a car when needed. This mindset has implications for financial establishments and drives decision-makers to invest in this new trend.

Integrated mobility platforms bring together all the options available for the commuter when planning a trip from an origin point to a destination point. Several transportation providers can use the single integrated smartphone app or on a web-page platform to sell their service to the prospective travelers. In some cities the services are bundled together and sold on a monthly subscription basis, so the user pays the transportation bill like a cable or telephone bill. This type of configuration has been termed as "mobility as a service," or Maas.

With these changes in the smart transportation system, legacy operators must develop a new strategy to remain competitive. Some options are to expand existing services and secondly to develop and offer new services at competitive prices.

References

1. Reuter, T. K. (2020). Smart City visions and human rights: Do they go together? In *Understanding the impact of technology on urban life. Harvard Kennedy School Carr Center Human Rights and Technology Fellow Director*. Institute for Human Rights/University of Alabama at Birmingham Spring.
2. Echendu, A. J., & Peter Claver, C. O. (2021). Smart city technology: A potential solution to Africa's growing population and rapid urbanization? *Development Studies Research, 8*(1), 82–93. https://doi.org/10.1080/21665095.2021.1894963
3. Kincho H. L., Jerome P. L., (2023). *Smart City: Technologies and challenges*. Retrieved from the world wide web. https://par.nsf.gov/servlets/purl/10126466
4. Graham, D. P. (2013). *A comparative evaluation of FDSA, GA, and SA non-linear programming algorithms and development of system-optimal methodology for dynamic pricing on I-95 Express*. Doctoral Dissertation. www.etd.fcla.ucf.edu

Chapter 5
Smart City Innovations in Integrated Payment Technologies

5.1 Benefits and Threats Related to the Implementation of the Smart City Payment Systems

As mentioned in prior chapters, the smart city planner will likely encounter challenges to development in a smart city project. Some of these challenges include:

1. Lack of technical knowledge and data network expertise.
2. Individuals and companies who are used to doing things in the old traditional way and don't want to change.
3. Companies or teams of companies working in silos and unwilling to share information or expertise when working on the same project.
4. Since the smart city collects a large amount of data on transactions from individual people, there is a risk that this data could be used for the wrong purpose, so the issue of privacy is one which becomes a big challenge when developing a new smart city project.

In addition to challenges or threats in the implementation of a smart city, there are several major benefits.

1. The first and foremost obvious benefit is the data-driven decisions that can be made to allocate resources in the best most optimal way.
2. Along with using data to make data-driven decisions, another primary purpose of the data is to use it to provide forecasts and trends for population travel and energy usage during peak hours of the day.
3. One common phenomenon that has emerged in smart city development is the concept of a public- private partnership. This type of partnership is one where a governmental organization or a city administration would partner with a corporate entity to help to develop a smart city capability search as a network or a transaction processing system which is built for the services of the city population.

© The Author(s), under exclusive license to Springer Nature Switzerland AG 2023

D. Graham, *Integrated Electronic Payment Technologies for Smart Cities*,
https://doi.org/10.1007/978-3-031-38222-2_5

4. Of utmost importance in a smart city is the ability to save money and resources and provide the maximum benefit at the lowest cost. This is one major benefit of developing a smart city project.

5.2 Smart Infrastructure

Smart cities must have smart infrastructure. In order for the successful implementation of smart cities, we must have the backbone in place to support the effective operations of the smart city. So, this includes smart networks that are interconnected spatially and also smart devices which are able to operate on the smart communication network. Smart city must also have smart mobility infrastructure. The smart mobility infrastructure includes electric vehicle charging systems widely available, as well as driverless cars and shuttle buses for mass transit. The mass transit system must have wired as well as wireless connections for online payment of services. This includes smart cards; mobile wallet payments such as Apple Pay, Google Pay, and Fitbit pay; as well as credit and debit card payment online. Along with smart mobility, the smart city must also have infrastructure for smart energy systems. Thus, the smart city will be able to handle a distributed energy supply coming from individual homes and office buildings which have their own solar installation providing power not only for their own building but also for the grid [1] (see Woetzel, SMART CITIES: Digital solutions for a more livable future McKinsey Global Institute 2018).

This smart city must also consider smart water supply systems and smart wastewater systems to make more efficient ways of providing these necessary services for its growing population. In many smart cities, solid waste collection has been streamlined to where the collection agency is able to monitor the level of waste in residential and commercial neighborhoods and schedule the pickup when the time is optimal and when it is necessary to pick up and dispose of this waste. Housing in smart cities is to be designed so that they are demographic and socioeconomically appropriate for the residents. This is important since we need to provide access ability for those who are not fully mobile physically, and we need to also consider gender when designing buildings as well as age and other factors.

The main infrastructures of a smart city consist of [2]:

1. Transportation infrastructure
2. Energy infrastructure
3. Housing infrastructure
4. Waste infrastructure
5. Communications infrastructure

As previously mentioned, transportation for structure includes all the buses, the trains for mass transit as well as the vehicle for ridesharing and for private transportation shuttle buses which are driverless, and air transportation and freight transportation. Some of the structures that must be in place to facilitate these transportation types include rail stations, subway station, bus stations, airports, monorail systems, and the corresponding Roland facilities subway tunnels and the train tracks which must be built or already in place for the usage of these different modes of transport.

The housing infrastructure in a smart city consists of gated residential communities, apartment buildings, single family homes, parks, and areas used for industrial as well as commercial activities which are often combined buildings to include living quarters and stores together in one unit.

The entire energy infrastructure consists of the electricity, the gas that is transported into the city using pipelines from other cities or other parts of the country, so this must be set up to receive and distribute this gas that is purchased outside. The electricity must be carried through high-voltage lines, and transmission corridors with transformers capable of handling huge amounts of power to distribute to customers in the urban environment. The smart city also has facilities for generation of electricity by individual customers using solar panels and generators which can add energy to the electricity grid.

Telecommunication infrastructure consists of cabling the physical electronic components as well as any logical connectors which must be present in order to facilitate the large number of transactions as well as provide information to the citizens for their usage in planning trips as well as other city services.

The water infrastructure is very important to a smart city as it serves the purpose of providing clean drinking water as well as being able to handle the removal of wastewater from the residents and commercial buildings in the smart city, thus completing the cycle.

We will see that all of these infrastructures in the smart city must be planned in such a way as Toby coordinated with each other and in some cases use the same tunnel or physical facilities to move this service from a central location outside the city to the customers inside the city or urban region.

5.2.1 Smart Buildings

Smart buildings are a big feature in any smart city. A smart building will be energy efficient and also made from sustainable materials and sustainable designs and will be environmentally friendly.

The smart building is constructed of materials which are able to handle extreme weather such as extreme heat and extreme cold. Materials such as brick are often preferred in cold climates where it snows due to its ability to keep the heat inside the home and perform better than concrete buildings in these types of environments.

This smart building can be constructed in such a way as to provide the maximum amount of natural sunlight during the day, thus reducing the need for interior lighting and saving energy in this way. Smart buildings can also be equipped with smart meters which are connected to the grid and can indicate the demand for more or less energy at any point in time, thus managing the amount of energy usage throughout the day and through peak hours so as to minimize the total energy costs in the building. The smart building is also equipped with connected locks for the door so as to prevent crime or suspicious activities. Smart buildings will be equipped with environmentally friendly designs and green building materials which will provide a more comfortable and sustainable environment inside the building.

5.3 Smart City Roles

A smart city implementation project will include personnel in several different roles:

1. The smart city manager
2. The urban planner
3. The Internet of Things designer
4. The data center manager

5.4 Payments Through Chat Bots and Messengers

What is a chat bot? A chat bot is a software application that is specifically designed to collaborate and understand human interactions responding with text and voice [3]. Over the last decade, several major companies have employed chat bots to help customers perform transactions online as well as to provide general instructions in completing forms and doing business with their companies. Some of these chatbots include animations such as Siri and Amazon Alexa as well as Microsoft Cortana.

5.5 Smart City Transportation System

Central to the concept office smart city is the smart transportation system. This will consist of smart services and ability to plan and execute trips with multiple modes of transportation search such as train the bus, or ride-share to subway, for example, while moving between different modes of transportation seamlessly.

In order to enable the seamless transfer from one mode or transportation to another mode, the use of a smart card and a connected network is vital.

Also of importance is the deployment of electric vehicles, ridesharing vehicles, and bikes, which will reduce both congestion and the pollution levels in the urban smart city.

5.6 Payment Systems in Smart City

The payment system in the smart city takes advantage of new technologies such as mobile wallets, RFID-based tag systems for automatic tool collection, credit and debit cards, and the aforementioned messenger-based payment systems [4].

5.7 What Are Integrated Payment Systems and Their Importance

An integrated payment system is the last piece of a seamless smart economy or city. These tools accept a variety of payments so that consumers have the option of paying through their preferred method. When an integrated payment system works in conjunction with an e-commerce platform, consumers are offered a much more personalized shopping experience.

This system allows an order to be tracked (or transaction) made on an e-commerce website. Each order is mapped to its respective payment ID, allowing the matching of receipts with invoices automatically. This eliminates the need to go back and check order details in a business database.

This is why an integrated payment system works better for a smart city which can also be called a payment gateway.

The integrated payment system works more like a payment application that keeps track of all transactions at every outlet. Here is a simple illustration. Saving receipts of all spendings at different outlets during shopping helps to keep proper records of all transactions. At a later time, these stored receipts can be cross-checked at home to a payment application. That is exactly how the integrated payment system does except this time; it is an automated operation in the form of a payment application. Aside from the quite obvious benefits of saving time and resources, chances of human error are eliminated, and the automated process provides tailored payment analytics.

5.7.1 Driverless Shuttle at University of Florida

The University of Florida Transportation Institute has partnered with the city of Gainesville and the Department of Transportation to develop a driverless shuttle for the downtown Gainesville area.

The shuttle transits between the University of Florida to the downtown area making six stops along the way. The (UF) driverless shuttle program began in 2020 with free rides for passengers who want to try out the vehicle. Although the shuttle operates autonomously, there will be a conductor onboard at all times to review and override any actions taken by the master computer which controls the shuttle. There are some distinct conditions which the shuttle may need a human to intervene to make the right decision. One example which the author observed was when there is construction in the path of the vehicle; this can sometimes cause the shuttle to stop and not have a computerized option to proceed. In this case the conductor needs to intervene and continue the motion of the shuttle while avoiding the construction and any oncoming traffic.

There are a total of six mounted cameras on the university shuttle which are used for monitoring the traffic and inside of the vehicle. According to the conductor on the shuttle, the feed from these cameras is live streamed to the University of Florida and Department of Transportation.

When passengers want to get off the shuttle, they press a “Request Stop” button inside the cab which signals the system to stop at the next available stop. The shuttle is operated on the streets of Gainesville at a speed 9 miles/hour but has the ability of speeds exceeding 25 mph. The shuttle connects via Wi-Fi to connect to the traffic signal and read the color of the traffic lights at a signalized intersection. The shuttle operators are provided by a company called Transdev based in France.

The UF driverless shuttle stopped operation for a period of time due to another similar system in another state having a minor accident. Due to this, all driverless shuttles carrying passengers onboard stopped operating until the problem was resolved. After the problem (which was due to sharp braking) was resolved, the UF driverless shuttle resumed operation with periodic inspections (www.tinyurl.com/gainesvilleshuttle) (Figs. 5.1 and 5.2).

Fig. 5.1 UF/Gainesville driverless shuttle

Fig. 5.2 Driverless shuttle RTS-Gainesville, FL

References

1. Jonathan W., Jaana R., Brodie B. et al. (2018). *Smart Cities: Digital solutions for a more livable future McKinsey global institute*. Retrieved front the world wide web. https://www.mckinsey.com/~/media/McKinsey/Industries/Public%20and%20Social%20Sector/Our%20Insights/Smart%20cities%20Digital%20solutions%20for%20a%20more%20livable%20future/MGI-Smart-Cities-Executive-summary.pdf
2. Al-Hader, M., & Rodzi, A. (2009). The smart city infrastructure development & monitoring. *Theoretical and Empirical Researches in Urban Management, 2*(11), 87–94. SourceRePEc.
3. Sengupta, R., & Lakshman, S. (2017). *Conversational Chatbots – Let's chat*. Deloitte.
4. Adam, M., Wessel, M., & Benlian, A. (2021). AI-based chatbots in customer service and their effects on user compliance. *Electron Markets, 31*, 427–445. https://doi.org/10.1007/s12525-020-00414-7

Chapter 6
The Use of Integrated Payment Technologies in Smart City Transportation Pricing Strategies

6.1 Introduction

As we have seen from previous chapters, integrated payment technologies can be vital to the efficient planning of a smart city's transportation network. It makes all the difference when commuters have the ease and convenience of a coordinated technological system and streamlined transaction processing.

The coordination of different modes of transport and payment systems leads to less confusion and greater clarity in the daily commute for millions of smart city dwellers.

Central to the concept of smart cities is the pricing strategy for optimal flow of traffic in the network. This optimal flow of traffic results in reduced travel time and overall reduced costs in the transportation network. In this chapter we will examine some common pricing strategies and how integrated payment technologies factor into optimal pricing strategies.

6.2 Highway Transportation Pricing Strategy

As previously discussed, the transportation pricing strategy that is used by many agencies is one which uses a lookup table to determine the rate for different levels of occupancy on the highway.

For instance, the I-95 managed lanes in South Florida have used a lookup table which calculates ($0.25) price increases for incremental changes in the travel demand as measured by the flow of traffic over a 15-minute period.

The location of the highway plays a significant role in the type of pricing strategy adopted; as we can see from table below, there are some strategies which are more

© The Author(s), under exclusive license to Springer Nature Switzerland AG 2023
D. Graham, *Integrated Electronic Payment Technologies for Smart Cities*,
https://doi.org/10.1007/978-3-031-38222-2_6

useful for downtown areas in an urban location, while a multi-lane highway may be better suited to another type of pricing structure.

Other major factors to consider include the demographics of the location, that is, what type of commuter is typical? What is the average value of time (VOT) for the traveler? Where are the popular, high-demand destinations? Thus, we see that a pricing strategy is highly dependent on several factors which must be included in the model and a customized approach must be developed for each urban location.

Pricing strategy is also paired with technology as can be seen from Table 6.1, the type of payment technology used for collecting the toll.

Thus, on a major highway, where vehicles are traveling at a high rate of speed and it is inconvenient to stop at toll plazas, a typical payment technology would be an overhead reader to read a RFID tag such as SunPass or E-ZPass. Recently, large portions of the Florida Turnpike have cut out toll collection plazas in the south Florida area. Instead of toll plazas and toll collectors, the toll is now collected using either:

1. Toll by plate
2. SunPass

These methods of payment do not require the commuter to stop to pay toll and saves on travel time.

Removing the toll booths and the toll collection plazas, however, also removes the need for toll collectors.

The table shown below (Table 6.1) lists some implementations of highway dynamic pricing on managed lanes both for domestic and international locations and the technologies utilized [1].

In the table above, the type of payment technology depends on the type of roadway and location.

Table 6.1 Pricing strategy on a sample of managed lanes

Road/Highway	Pricing strategy	Technology
SR 91 San Diego	*Express lanes*	*Auto*
I-95 Miami, FL	*HOT lane*	*Auto*
I-85 Atlanta, GA	*HOT lane*	*Auto*
I-10, Houston TX	*Express lanes*	*Auto*
I-394, Minneapolis	*HOT lane*	*Auto*
I-495, Washington, DC	*HOT lane*	*Auto*
I-25 Denver, CO	*Express lanes/time of day*	*Auto/license plate*
Fort Myers bridge, FL	*Variable price road*	*Auto/toll plaza*
SR 520 Seattle, WA	*Variable price road*	*Auto/license plate*
SR 167 Seattle, WA	*Variable price road*	*Auto*
Stockholm City Center	*Cordon/TOD*	*Auto/license plate*
Singapore City Center	*Cordon/TOD*	*Auto/license plate*

Thus, a two-lane rural roadway with a bridge may use one type of technology for payment, and a six-lane highway with a high-occupancy toll (HOT) lane may use a different type of payment technology.

The type of payment technology for road pricing on the roadways listed includes:

- Auto
- License plate
- Toll plaza/toll booth

The auto payment technology refers to automatic payments through the use of a RFID transponder. As the commuter drives past a toll gantry, the transponder in the vehicle is read by a radiofrequency ID (RFID) overhead reader that is constructed across the highway. The reader is able to collect the information on the RFID tag on the commuter vehicle (most often the front windshield) and obtain account information instantaneously. The toll or cost to use the managed lane is then deducted from the account based on the pricing scheme in use on the highway at that location.

Toll by plate is a type of payment technology that allows for the commuter to drive along a roadway and be able to pay the toll using electronic payment technology. High-definition cameras which are able to capture the license plate information of a vehicle driving at high speeds are mounted on poles or a toll gantry along the roadway. A picture of the license plate is taken which provides information about the owner of the vehicle, and a bill is sent out to the owner for the dynamic price or toll for using that section of the roadway.

Toll plaza is the least technologically advanced method of payment for the roadways listed. The toll plaza does not require electronic toll collection but uses toll booths with live toll collectors to collect cash payments for the toll. The toll plaza requires the traveler to stop and break their trip to pay the toll and thus represents a delay in time for the traveler.

The payment technologies listed in the table are by no means all the methods of payment which are available. These are a subset of a larger group of payment systems which apply to road pricing. In addition, there are other payment technologies for mass transit such as credit/debit card and mobile payments with a smartphone.

The innovation here is that all these payment technologies can be integrated on a single platform in a smart city for a smooth travel experience for the commuter. The commuter in the smart city will be able to pay for travel on the urban roadway and then change modes to mass transit or shared ride while using the same platform to pay for the entire trip.

The integration of the different modes of travel from different vendors onto one platform will allow the smart city administrator to see overall trip data. Trends for travel between popular origin-destination pairs can then be used for planning the resources to make these trips faster and more efficient.

This type of problem is well suited for a machine learning tool to automate the data mining and provide meaningful results. The system should be knowledgeable enough to recommend alternative routes on any leg of the trip so as to reduce travel time and congestion.

The data generated from an integrated travel system is of value to both system administrators and private vendors who may be planning services for the commuter. The question of who owns the data should be determined by a regulatory body. The data can be sold to providers, and privacy issues must also be taken into consideration as to what type of information is shareable and for what purpose it is used.

In one pilot program in the US state of Oregon, commuters were provided with an electronic dongle—a device which plugs into the OBD port in the vehicle. The device logs the travel miles in toll roads based on GPS technology.

Once the monthly billing period comes to an end, the commuter is presented with a bill for the miles driven and recorded on the device. His type of payment technology is known as "pay as you go."

In some European cities, global positioning system (GPS) is installed on commercial trucks which use the toll roads. The GPS devices logs the time spent and miles driven on each toll road. The owner of the truck is then billed at periodic intervals for the usage of the toll road.

For highway pricing strategy in a smart city, we can see that technology use is linked to pricing. Several.

Implementations of managed lanes utilize automatic payments through radiofrequency ID (RFID) technology. This technology facilitates simple, convenient payment of tolls without the need for the commuter to stop. The toll is electronically deducted from the user's account which is preestablished on a transponder. The flow of traffic is continuous due to no toll plaza or queueing on the managed lane, thus reducing congestion.

As mentioned in Chap. 3, the RFID reader transmits this account information to a remote computer using wireless technology and protocol typically using an electronic gateway. The computer then processes the transaction by deducting the current congestion price and updates the user account.

In many managed lanes, a high-speed camera is utilized. When employed in conjunction with the RFID reader, the camera can identify license plate information for commuters who did not have a valid account while using the managed lane. This allows the transportation agency to send those toll violators a bill to make payment by mail or otherwise.

Managed lane pricing and billing can be performed with the camera only. This method utilizes a toll-by-plate pricing strategy. The commuter(s) passes under a toll gantry equipped with high-speed cameras which can read license plate information in any weather condition at any time of day. These smart cameras transmit this information to a computer which has software and image processing algorithms used to identify the alpha-numeric characters on the image taken by the camera.

Once the license plate information is processed, the owner of the vehicle can be identified and a bill sent out for the toll (or congestion price) to be paid. In the same manner, violators can be identified and handled.

Another strategy in use is the pay as you use. In this method, users are equipped with a technology device that records their usage of roadways by miles, and at the end of the period, they make payment for the miles driven on the highway. This strategy has been implemented in the state of Oregon.

Smart city road pricing strategies are devised based on the stated objectives of the administration.

In developing the smart city strategy, several objectives may be considered:

1. Maximizing revenue on the Express lanes
2. Maximizing throughput on managed lanes
3. Minimizing travel time on managed lanes
4. Minimizing delay

These objectives can be combined with other constraints to determine the optimal pricing scheme which changes dynamically over a specified time period.

Once the objective(s) and constraints are determined, the optimal pricing schema can be obtained using established mathematical programming methods known as linear or nonlinear programming. These methods are able to handle multi-objective problems and nonlinearity in both the objective function and the constraints.

Later in this book, we will apply these methods to a case study example of the managed lanes on I-95 in Miami, Florida. The solution to the optimization problem may be developed using heuristic algorithms such as genetic algorithms, simulated annealing algorithms, and stochastic approximation algorithms.

6.3 Road Payment Technologies

Payment technologies data can be an excellent source of near real-time information on the travel patterns of the commuter. The data that is collected from current payment technologies is rich in its ability to break down commuter trips by frequency, destination, route choice, and mode choice.

Data collected from highway payment systems such as electronic toll collection systems can give insight into what times of day are peak hours for transportation and what demand volumes can be expected on the highway. This information is useful for forecasting future travel demand and utility of each corridor.

The demand forecast will be used to drive the pricing strategy for the roadway and determine when tolls on the managed lanes should be increased or decreased based on a dynamic pricing model.

As the commuter uses the managed lane to travel from origin to destination, the transponder is interrogated by the RFID reader on the highway. This information about the user's account is transmitted by the reader to an electronic gateway which collects data from multiple sources via a communications link between the reader and gateway. The data is sent by wireless communication protocols to the remote datacenter to a cloud computer system [2].

As computers process the transaction for each commuter, usage characteristics are developed for the arterial based on the number of transactions per unit time.

These transactional data are granular enough to provide a forecast of demand volume over time for the future.

Since the managed lanes are priced based on demand for use, we can adjust the pricing on a near real-time basis due to the use of the auto-RFID technology capabilities of the system.

Forecasting demand volumes are just one advantage of this auto-RFID technology. One can gain useful insight into the popular destination point for most commuters when the highway managed lanes are subdivided using segments. If most of the traffic has the same destination during a certain period, then the pricing strategy can take this information into account when planning a dynamic price for the corridor segments.

Another key insight provided by the payment technology is the mode of transportation for traveling the managed lanes. Mass transit buses using the managed lane corridor can have a significant impact on the capacity of the corridor, and ridership on the mass transit can be an important factor when developing a pricing strategy for a managed lane such as a HOT lane or Express lane system.

Besides mode of transport and demand volumes, route choice decisions for each origin to destination pair can also be deduced and that information incorporated into the pricing strategy by the smart city planners when this data has been collected over a sufficient time frame for it to be used reliably in determining demand and price.

As mentioned in Chap. 3, smartphone applications have been developed for managed lanes in several states. Peach Pass [1] is one example of the I-85 in Georgia. In addition to using smartphone applications for highway electronic toll collection, smartphone apps have been developed for parking, ride sharing, and mass transit as a payment technology. When these software applications are implemented as a payment technology, the payment transaction can be integrated with blockchain and cryptocurrency for a more secure system.

Payment for parking space in urban areas is simplified with smartphone applications. The application provides real-time information on the utilization of parking space capacity and the demand volumes at different times and different locations around the city. The pricing for these parking spaces can be dynamically adjusted based on the estimated forecast for demand. Smartphone applications for parking also can provide location services for open spots. When using an app for parking payment, the location of the spot which was paid for can be registered on the system as being reserved and thus it is unavailable for the time paid for. Smartphone apps for parking facilitate the process of extending the time of parking at a spot. Instead of needing to physically return to the parking spot to make payment with coins or credit card, the user may simply pay on the smartphone and avoid the need to interrupt his/her business.

One of the most direct areas of application for the mobile app payments is that of ride sharing. Several ridesharing companies such as Uber and Lyft have grown in magnitude with the progressive use of the smartphone and convenience of the apps which have been developed for it.

Smartphone technology when used for ridesharing can provide invaluable information about rider demands during different times of the day. This data can be aggregated and analyzed to drive pricing strategy for these companies.

It is known that these ridesharing companies use "surge pricing" which takes the demand volume during peak hours to calculate a premium price for service in congested regions. In addition to time fluctuations of demand being used to determine pricing, the destination can also play a role in a pricing strategy. Airport rides and special events are prime high-demand locations which individuals are willing to pay a premium to arrive on time [3].

Payment platforms for rides on ride-sharing apps can be integrated with blockchain technology to provide added safety for the transaction process. Blockchain uses a distributed ledger system which is difficult to infiltrate and tends to be tamper resistant. Some smart city programs have already developed blockchain-enabled payment platforms for all modes for transportation including mass transit, car-sharing, parking, and managed lane tolls [3].

Once this smartphone app data is analyzed, many trends can be identified using data science algorithms, and optimization methods can be applied to help reduce travel times and provide a more efficient and reliable transportation service for urban customers.

Big data from smartphone apps can also be applied to provide customized service for different user groups in the region. For instance, more trains can be put in service for high traffic areas such as a university or a city hall area.

6.4 Electronic Ticketing Technology for Mass Transit, Ride Sharing, and Parking

As mentioned in previous chapters, technologies for payment of fares on urban mass transit (subway, bus) include magnetic stripe cards which can be read by an electronic reader at the point of sale. Such cards provide ease of use and do not require the presence of a fare collector at the station. This results in lower cost systemwide and higher efficiency which many city transit operators use as a strategy to manage budgets and improve profitability.

In addition to the magnetic stripe cards, there are contactless systems in which the reader is able to interrogate the card without the need to swipe. This can be more efficient since the card stripe is not damaged by repeated swiping [3].

For many transit agencies, commuters must purchase the card for their specific system. However, major credit card companies have initiatives which allow direct fare payment using their credit card with no need for users to buy the magnetic stripe card for that system. This eliminates a step in the process and is a consideration which practitioners must decide if they want to adopt for their system.

Mass transit fares can also be prepaid using web applications and presented at subway stations using contactless readers. Electronic wallets (e-wallets) allow commuters to pay for services including transit without purchasing a card or paying a representative at the station. Such e-wallet systems such as Apple Pay and Google Pay are already being used widely in other industries and are being applied to mass transit as well.

Ride-sharing companies have to a large extent built their companies on the smartphone technology which makes it easier for customers to order service. Since the ride-sharing company has real-time knowledge of demand in each region/city, they can apply dynamic pricing algorithms which help them to increase revenue during peak periods. For instance, on one recent trip using a ride-sharing company from the airport to hotel, the price was listed at $18. However, during the return trip to the airport from the same hotel at peak period, the cost was $20. This was due to premium pricing for the higher demand at peak period, thus increasing revenue for the same distance driven.

In recent years, Internet technology has also revolutionized the parking industry. Previously, car parking lots and urban areas would use parking meters to charge users to park their cars in a parking space for a time period. With e-commerce on smartphones, new apps have been developed which allow commuters in dense urban regions to pre-book and pay for parking electronically. Such apps include Park Spot and ParkMobile which is covered in greater detail later in the book [3].

6.5 Static vs. Dynamic Pricing

Pricing for highways can be done using static or dynamic methods. By static we mean that the price does not change with the conditions of the traffic but stays the same over a period of time. On the other hand, dynamic prices change with the changing volume of traffic present. Examples of static pricing can be seen on a toll road such as a turnpike, where the cost to use the turnpike is constant throughout the day regardless of the time of day or volume of traffic on the road. On some highway facilities, a type of congestion pricing is used which is based on the time of day. This is called time of day (TOD) pricing. As mentioned, the price to use the facility is set at a fixed value at certain times of the day. This value does not change based on the volume of traffic on the highway but on the peak period for that region. Thus, as an example, a fixed price is set for the morning hour commute (6–9 am), and a fixed price is set for the evening commute (4–7 pm). Since static pricing sets a fixed price for the use of the facility, it does not explicitly use price elasticity to manage traffic volume on the transportation facility. The concept of price-elasticity is one which measures how demand for a service (such as a transportation facility) changes as the price of that service or product changes. Dynamic pricing involves setting incremental price levels for a service or product and changing the price by increments with the change in demand for that product. With dynamic pricing, since the price changes with time, the dynamic pricing algorithm will review the demand at preset time intervals to determine if an increase or decrease in price is warrantied and if so, how much of a change should be made. As an example, from transportation modeling, in the case of the I-95 Express lanes, the increment is set at 25 cents, and every 15 minutes the traffic density is reviewed to determine if a change is necessary. This review is based on a feedback mechanism which can measure the actual traffic flow on the highway HOT lane and feed this value into a computer system. The computer algorithm then sets a threshold demand level at which the price will increase to the next increment.

Variable pricing in Lee county was implemented in 1998. The purpose of this program was to reduce congestion during the peak AM morning and evening hours on the Midpoint Bridge and the Cape Coral Bridge. The terms of the program were to allow for a 25% discount in the toll rate in the hour before and after peak AM hour and the hour before and after the peak PM hour. These time periods are referred to as the "shoulder hours." By reducing the rate at these times, drivers are attracted to change their departure time in exchange for a lower toll. Since the peak AM hour are 7–9 am, the morning shoulder times are 6:30–7 am and 9–11 am. The evening shoulder times are 2–4 pm and 6:30–7:00 pm. Because the Reduced Fare Program can be used in addition to the variable price program, it is possible for commuters to pay only 75 cents for the one-way toll when using the Cape Coral causeway and Midpoint bridges around the daily peak period. It is important to point out that the toll amount does not fluctuate with traffic volume on either of these bridges and that the price does not increase during the peak period due to these programs.

6.6 Bridge Pricing in Lee County Florida

The Ft. Myers Florida metropolitan region has three bridges which cross over to Cape Coral and Sanibel Island. These bridges, which use some form of congestion pricing, are:

1. Cape Coral Bridge
2. Midpoint Bridge
3. Sanibel Island Bridge

6.7 The Midpoint Bridge

The Midpoint Bridge is the most northerly of the three bridges. The Midpoint Memorial Bridge as well as the Cape Coral Bridge spans the Caloosahatchee River. The Sanibel Island Causeway Bridge connects Ft. Myers and Sanibel Island over the Gulf of Mexico. The Midpoint Bridge was completed in 1997 and features one-way tolling with electronic toll collection. Only westbound traffic is tolled and users with a SunPass transponder are allowed on an exclusive lane where "open-road tolling" is in effect. Open road tolling refers to the ability of users of SunPass transponder accounts to pay the toll without having to stop. SunPass users of the

Lee County bridges can use the electronic toll collection system in North Carolina due to an interoperability agreement between the two transportation agencies. Similarly, North Carolina drivers who possess a Quick Pass transponder for use on North Carolina toll roads can use their transponder on the bridges in Lee County without alteration. Level of service (LOS) on the Midpoint bridge toll road is expected to be at a LOS value of "D," on the scale of A–F. With level of service "D," the bridge can be expected to accommodate 3240 vehicles per hour in each direction (both eastbound and westbound) at a speed of 50 mph. Traffic peaks on the Midpoint Bridge during the weekday between the hours of 4 pm and 6 pm on the westbound lanes. On weekend days, the traffic gradually increases during the day and maximum flow occurs between the hours of 12 pm and 6 pm, after which it declines, in both directions. The toll on the midpoint bridge is $2.00 for cash payment and $1.50 for users with a SunPass account.

6.8 Sanibel Island Causeway

The Sanibel Island Bridge was opened in 1963. The bridge consists of three spans with manmade islands constructed in the Caloosahatchee River. The bridge is designed to accommodate a capacity of 1200vph in each direction with an average speed of 30mph. According to the Highway Capacity manual, a bridge with these operational characteristics would represent a level of service of "D." As with the Midpoint Bridge and the Cape Coral Bridge, only the two lanes westbound on the Sanibel Island are tolled. The toll is collected at the exit of the bridge via a toll plaza with an electronic toll collection (ETC) lane for SunPass users as well as two cash lanes.

6.9 Cape Coral Bridge

The Cape Coral Bridge is located south of the Midpoint Bridge and north of the Sanibel Causeway. The Cape Coral Bridge opened in 1964 and was the first bridge across the Caloosahatchee River to connect Ft. Myers to Cape Coral. Like the Midpoint Bridge, tolls are only collected in the westbound direction. Usage characteristics are nearly identical to the Midpoint Bridge, except for the time period when bridge lanes closed for construction. The bridge was designed to have a capacity of 3240vph in each direction with speeds of 50 mph. The level of service (LOS) on the bridge is designed to be LOS "D." The peak hours of operation and demand are quite similar to the Midpoint Bridge, in that the westbound traffic during the hours of 4 pm and 6 pm represent the highest percentage of volume flow daily. In addition, the weekend traffic shows a traffic profile where the hours from 12 pm and 6 pm have the highest volume. The Cape Coral Bridge is located south of the Midpoint Bridge and north of the Sanibel Causeway.

6.10 National Interoperability

Electronic toll collection on the State Route 91 is done using a transponder. This transponder is called "FastTrack" and allows drivers on SR 91 to use the toll collection systems anywhere; tolls are collected in California. Due to the national movement to make electronic toll collection easier, the federal agencies have developed a protocol called "MAP 21" which is used to standardize transponders nationwide. For instance, drivers in Florida should be able to use their SunPass transponder in California or at the Northeastern states electronic toll collection systems. The IBTTA (International Bridge, Tunnel and Turnpike Association) is working with several state transportation agencies to attain nationwide interoperability of all transponders. A 2017 date for implementation of this national interoperability is set as a goal. The SR 91 Express currently has only one entrance and one exit. When the current extension into Riverside County is completed, the county line will provide another entrance and exit location for the expressway. Thus, the driver will have the chance to leave the toll road halfway, before the additional toll is charged to their transponder. Although there is no toll booth at the county line, there will be another transponder reader to perform the toll transactions going into Riverside County which will allow the driver to continue their trip without stopping.

6.11 Radiofrequency Identification (RFID)

Radiofrequency identification (RFID) is a technology which is used widely commercially for supply chain networks. Some companies which use RFID include IBM, Walmart, Intel, Sun Microsystems, Department of Defense, as well as Microsoft Corporation. In a supply chain, RFID can be used to automatically log products and materials as they move through the supply chain network. RFID can reduce out-of-stock conditions and improve efficiency in inventory systems. RFID can be used to improve forecast accuracy. Due to the many advantages of RFID for transaction processing, several transportation agencies have adopted it for use in electronic toll collection. Thus, some of the RFID-transponder technology systems created include FastTrack (California), E-ZPass, and SunPass (Florida). Each of these transponder technologies uses a different frequency block on the frequency spectrum and is not compatible with each other.

The RFID transponder tag is placed on the inside of the vehicle windshield and reads as the vehicle drives under the antennae placed at the entrance of the toll collection facility. The correct amount of the toll or congestion price is deducted from the prepaid account of the vehicle owner or operator. The RFID tag in the vehicle can be read even at high vehicle speeds. If a vehicle is detected but the RFID tag is not read, the system will automatically take a photograph of the license plate and contact the owner of the vehicle to enforce the fine. If the owner of the vehicle possesses a tag, then the system will only deduct the toll and not the penalty.

Radiofrequency identification tags work by emitting and receiving low-frequency radio waves. The device which emits the radio signal is called the reader and the device from which information is to be read is called the tag. RFID chips were first introduced in the USA during World War 2, when they were used to identify aircraft from the USAF. It helped Air Force pilots to identify which airplanes were US Air Force planes and which were enemy planes. Since then, it has grown tremendously in its applications and technology. RFID chips can be as small as a square centimeter or less and are subdivided into passive devices and active devices. Passive RFID devices have no power source of their own and are only activated when the reader sends a radio wave signal to read the device. RFID allows for the transmitting of data from the source without any physical contact. The data read by the reader can be transferred directly to a computer for transaction processing. Unlike bar codes, the RFID tag, or transponder, can be read from a range of several hundred feet and can read multiple tags at once. In addition, RFID tags do not require an unimpeded line of sight in order to read the tag or transmit the data to the reader. Passive RFID tags consist of the following key components: integrated circuit, antennae, and substrate. The circuit is typically able to store 1–2 k bits of data. The integrated circuit chip can be a read-only (RO) chip, write-once (WO) chip, and read-write (RW) chip. The antennae are used to transmit data stored back to the reader, and the substrate is typically some substance which connects the circuit to the antennae. The performance of the tag depends on the size and shape of the antenna.

6.12 Electronic Toll Collection

Electronic toll collection (ETC) is a method of payment of road charges used on toll roads in the USA and internationally. With electronic toll payment, vehicles use a RFID transponder which is typically affixed to the front windshield of the vehicle. The account information and balance are encoded in the RFID tag (transponder) and can be communicated to a reader when the vehicle passes within the proximity of the antenna of the reader which is set up along the highway over the payment lanes. The reader generates an electromagnetic field which is capable of interrogating and reading RFID tags on the passing vehicles at a high rate of speed. The reader-antenna combination can also read multiple tags within its range at the same time. Once the vehicle passes the reader, the information on the tag is processed by the reader's computer, and the toll is taken from the tag holder's account. The transaction is also recorded on the RFID tag, and the information is displayed by indicator lights which show one of three states: (1) toll paid, (2) toll not paid violation, and (3) balance low. The advantage of using electronic toll collection is that drivers do not have to stop at a toll booth to pay the fee but can continue through the toll gantry at a high rate of speed. For congestion pricing facilities, this is critical. Toll plazas require vehicles to stop, form a queue, and get service to pay the toll. All of this requires increase in travel time and increase in congestion which counteracts the purpose of all congestion pricing facilities. Thus, electronic toll collection solves this congestion problem by allowing free-flow speeds through the HOT or Express lanes while processing the transaction and collecting the fee for the convenience of reduced travel time and increased speeds compared to the general purpose, non-tolled (free) lanes. Electronic toll collection is used on such high-occupancy toll (HOT) lanes as I-95 Express, in Miami, Florida (SunPass) and all throughout New England (NY, NJ, MA, PA) with E-ZPass transponder systems. Electronic toll collection eliminates the need for a brick-and-mortar structure to house toll booths. It also does not require the need for toll attendants, thus reducing the cost related to those factors in the overall fee structure. Radiofrequency identification systems are also used widely for parking lot payments and access to restricted areas such as facilities for security personnel and high value storage. Electronic tolling facilities report a 99.9% accuracy rate for reading tags and processing transactions correctly.

6.13 Case Study 2: Roy Selmon Expressway, Tampa, FL [4]

The Roy Selmon Expressway is a major highway in Tampa, Florida. It connects the outlying towns and communities to downtown Tampa and the business districts. The highway is a reversible road which allows peak hour traffic going into downtown Tampa the right of way during the morning hours and the traffic leaving downtown have the right of way during the evening hours. The expressway is tolled based with prices depending on the segment traversed and uses toll-by-plate technology for capturing license plates and billing the commuters for their trip on the expressway. The cameras set up on the roadway can read license plates at high speed without the need for vehicles to stop at a toll plaza.

6.14 Connected Vehicles on Selmon Expressway

The Selmon Expressway was included as part of the US Department of Transportation (dot.gov) Connected Vehicle (CV) Pilot program for Tampa, Florida (https://www.tampa-xway.com/wp-content/uploads/2018/07/TEX-497_Annual_Report_PRINT3-single.pdf). The expressway was outfitted with connected vehicle equipment which serves as the infrastructure for receiving and transmitting signals between vehicles driving on the road. The equipment can warn drivers who are participating in the connected vehicle pilot program of other wrong-way vehicles driving during certain hours and warn them if they are driving the wrong way on the expressway (Fig. 6.1).

The Selmon Expressway extends into downtown Tampa where it intersects with local roads. Smart city and connected vehicle technology installed on the infrastructure on downtown streets include:

Fig. 6.1 Toll gantry for toll-by-plate payments, Selmon Expressway

1. Roadside units
2. Pedestrian monitors
3. Smart LED streetlights

The roadside units are typically placed at intersections and mounted on traffic signals. Transit buses communicate with the roadside units and can obtain priority at traffic signals. This control of priority at the traffic signals for mass transit buses allows for reduced congestion in the downtown areas because they move large numbers of commuters at a much faster rate during the peak hour, coming into work in the morning and returning home in the evening (Fig. 6.2).

The Tampa connected vehicle pilot program also includes streetcars which run on rails through the city. The street cars are equipped with onboard units and antennas which mount atop the vehicle. The installed equipment will be able to detect the presence of vehicles crossing in front of the streetcar or crossing the tracks. In this event, a warning will be displayed at the front of the streetcar and allow for evasive action. The smart city pedestrian monitors are installed in the downtown area and monitor the presence of pedestrians on the road and in the crosswalk. This information is transmitted to connected vehicles nearby which can then use this knowledge to avoid collision with crossing pedestrians. The CV pilot thus enhances safety by reducing injury or fatalities through collisions (Fig. 6.3).

Smart LED streetlights are installed in the downtown areas to provide illumination and improve safety. The smart LED lights utilize a light sensor which detects

Fig. 6.2 Connected vehicle roadside unit at intersection

Fig. 6.3 Pedestrian monitor cameras, Tampa FL

ambient light intensity, and when that light intensity falls below a preset level, the LED will be activated, and the light will turn on. In a connected smart city, the LED can be controlled externally and set to any desired level (Fig. 6.4).

6.15 Summary

In this chapter we focus on the technology which enables smart city applications. The smart city relies on the use of novel technology to interface with the commuter and collect data. These technological innovations include highway RFID readers and transponders for electronic toll collection systems. It also includes the communications of these readers with the cloud computers and transaction processing at the data center.

Fig. 6.4 LED smart streetlight

The use of smartphone apps for integrated payment of mass transit is of increasing value for commuters as they plan trips in urban environments which often require multimodal transport to travel from origin to destination. The ease of use of the smartphone for payment combined with the security features of blockchain-distributed transaction processing provides a seamless and robust platform to assess and execute transportation alternatives.

As we have seen, there are several payment technologies used for payment on urban roadways in a smart city. One common issue with these payment methods is that they are not integrated with the other modes of transportation or the network.

The innovation of blockchain can provide integration of roadway payment with other modes of transportation within the smart city network.

Because blockchain is a distributed ledger system, it can integrate multiple vendors and modes to enhance security of transactions. There are many uses for blockchain including but not limited to:

1. Management of fraud
2. Smart contracts
3. Managing asset ownership
4. Voting transparency
5. System integration

Although blockchain is a system for performing transactions securely across a distributed network, it is not a form of currency. Cryptocurrency is often used as the currency for transactions on the blockchain.

One cryptocurrency is Bitcoin. Bitcoin has had highly volatile changes in values over the past several years. Cryptocurrency can be stored in a hard wallet or soft wallet. Hard crypto wallets are physical devices which can be connected to a single computer to store the existence and value of the crypto asset. A soft crypto wallet is an online location where the cryptocurrency value is stored and retrieved.

Blockchain also comes with its own set of challenges. The distributed ledger system requires a process called mining to validate transactions, and these processes can be energy consuming. The information stored on the blocks in blockchain are designed to be unchangeable; thus, if there is an inaccuracy on a block, it is very difficult to correct. In most cases experts agree it is advisable to create a new block to override the old block information which can't be changed.

Privacy and protection of sensitive information can also be an issue when transactions are performed on the blockchain as opposed to a centralized system with a single institution.

Regulation of how this data is stored and accessed within the smart city context must be considered before adoption of this system as an integrated platform.

In a later chapter, we will cover the blockchain and how it works in more detail, and we will also delve into the use of smart contracts to automate some financial functions and increase the efficiency of the system even greater.

References

1. Graham, D., (2013). *A comparative evaluation of FDSA, GA, and SA nonlinear programming algorithms and development of system-optimal methodology for dynamic pricing on I-95 express,* PhD dissertation University of Central Florida.
2. Ben-Akiva, M. E., & Lerman, S. R. (1985). *Discrete choice analysis: Theory and application to travel demand.* MIT Press.
3. Sattlegger, K., & Denk, U. (2014, March). *Navigating your way through the RFID jungle.* Texas Instruments.
4. Tampa Hillsborough Expressway Authority. *Driving innovation.* Implementing Change. https://www.tampa-xway.com/wp-content/uploads/2018/07/TEX-497_Annual_Report_PRINT3-single.pdf

Chapter 7
How Analytics Can Be Used to Guide Smart City Strategies

7.1 Introduction

Analytics has long been a vital tool in many areas of business and scientific work; however, in recent years, the use of data science and data analytics has increased exponentially. This is due largely to the increase in computing power and storage capacity which allows for the manipulation of huge amounts of data.

Big data analytics can provide valuable insight into what the consumer wants and does not want, what they are willing to pay for transportation services, and how much utility the consumer places on a service or goods.

Big data analytics can predict future trends in travel and smart city transportation. Both the mode of travel and peak travel demand can easily be captured with analytics techniques. Once the demand profile for a particular smart city region is captured using analytics, an appropriate pricing strategy can be developed to optimize pricing and minimize travel times.

We will begin this chapter by looking at some models which can be applied to highway managed lane pricing once the data is collected.

Several desktop software applications have been developed for data and statistical analysis of small- to medium-sized datasets. The data from mass transit and highway data collection systems consist of near real-time large volumes of streaming data which is best handled in a cloud data center environment.

These data have specialized data science algorithms which have been developed and optimized to quickly and efficiently manipulate the datasets to make actionable decisions in a short period of time.

Programming languages have been developed with specialized database features to handle these huge datasets efficiently. Some of these include Hadoop, Apache Spark, and Python Pandas.

When transportation data is sent to the cloud, the cloud service provider will have applications on the data center computers to process the data transactions.

© The Author(s), under exclusive license to Springer Nature Switzerland AG 2023

D. Graham, *Integrated Electronic Payment Technologies for Smart Cities*,
https://doi.org/10.1007/978-3-031-38222-2_7

Along with these applications, there are several applications which provide data mining operations, regression and forecasting, as well as optimization functions.

Machine learning on the cloud is offered through major cloud service providers. Machine learning can use the information collected from millions of Iot (Internet of Things) sensors and perform training and feature models to help develop predictive capabilities using the data gathered [1].

Pattern recognition, image recognition, and voice recognition are some main applications of machine learning which can be accomplished with large volumes of data from IoT networks. Because of the major advances in speed and accuracy of these technologies, privacy concerns have become a real issue when these machine learning algorithms are applied to the public forum on city streets. The use of public data to monitor, identify, and control individuals without their knowledge is an area in which legal experts are still working to catch up to the technology of today. Business intelligence for urban mass transit networks is also a great application for cloud services. Microsoft Azure and Amazon Web services both offer business analytic applications to work on the data stored on the cloud. The transit manager can perform queries and view results for different business questions from these analytics applications. Trends for bus and subway ridership can be used to forecast demand for peak-hour services into the future. The management of supply for this demand can be tailored by region in a city or by demographics such as the type or age of the riders. This level of insight is vital to smart city planners when working to reduce cost and optimize travel efficiency. For highways and downtown areas, pattern recognition can be useful in improving safety and reducing accidents at highly congested areas. Pedestrian safety can be greatly improved using IoT pedestrian monitoring cameras at traffic signals [3]. These machine learning algorithms can be trained to identify the conditions which lead to accidents and therefore make adjustments to avoid those conditions in the first place.

Structured queries can also be done on the data which is collected and sent to the cloud. The cloud data services such as Azure have SQL database features which are useful to analyze data in tabular form and extract results from SQL queries.

The insights gained from using this type of SQL queries are valuable in performing scenario analysis and answering what-if-type questions.

Python Pandas data frames can be useful in efficient manipulation of data and dissecting databases to provide business intelligence. Because of the popularity of the Python Pandas database, many programmers have used this programming language with cloud services to create new applications for specific or custom purposes.

Python and the R programming languages can also provide machine learning functions as well as implement regression forecasting models.

7.2 Problem Formulation [2]

Basic Assumptions In formulating the model, the following basic assumptions were used to facilitate the analysis:

1. Demand Flow *D* is known for each 15-min period.
2. Density is measured by loop detectors and updated every 15 min.
3. No queuing occurs on the Express lanes (we constrain the speed to be >45 mph).
4. The capacity does not change during the period ($Q = 2000$ vphpl).
5. The same number of vehicles exits any access point as it enters; thus, the flow along the corridor is the flow at the entry point. The nonlinear function minimization is carried out using MATLAB.

The goal of the 95 Express is to minimize the total travel time along the I-95 corridor for both the Express lanes (EL) and the general-purpose lanes (GP).

To express this mathematically, we will sum the travel times for *all* the demand flow entering the corridor:

Thus, total travel time =

$$\sum_{0}^{x1} t1(x1) + \sum_{0}^{x2} t2(x2) \tag{7.1}$$

We then substitute the travel time equations into the equation above to obtain the travel time optimization problem, including cell transmission constraints.

Since we are formulating the general model, we will consider a multi-period time horizon with time periods numbered from $k = 1$ to n. Each time period represents a 15 min interval on the HOT lanes.

The results are the optimal flow values for the Express lane (EL) and general-purpose lane (GP), which minimize the total travel time along the corridor given the constraints. Knowing the optimal flow values, we calculate the travel time along the Express lane. Since travel time is related to toll amount, we can calculate the appropriate toll value using the following equation [2, 3, 5]:

$$\text{Travel time}\, t = \text{TT} + \text{toll} / \text{VOT}$$

where TT = free–flow travel time

VOT = value of time. This value is usually taken to be \$15–25/h depending on location.

7.3 Optimization Results [2]

Results and Sample Calculations (Figs. 7.1, 7.2, 7.3, 7.4, 7.5 and Tables 7.1, 7.2, 7.3, 7.4, 7.5)

The following figures are the function minimization plots for demand values range 3000–14000 vph

$$t1 = \mathrm{TT} + 0.0625 \times \left((x-1) + \left((x-1)^\wedge 2 + \left(0.0016 \times (x) \right) \right)^\wedge 0.5 \right)$$

$$t1 - \mathrm{TT} = \mathrm{toll} / \mathrm{VOT} = 0.02134 \Longrightarrow \mathrm{toll} = 0.02134 \times 20 = \$0.4269 \Longrightarrow \mathrm{add.Toll} = \$0.50$$

$$\mathrm{Toll} = \$.50 + \$.50 = \$1.00$$

$$\mathrm{Demand} = 10000\,\mathrm{Cap}(\mathrm{EL}) = 4000, \mathrm{density} = 75\,\mathrm{vehicle} / \mathrm{mile}$$

$$t1 = \mathrm{TT} + 0.0625 \times \left((1-1) + \left((1-1)^\wedge 2 + \left(0.0016 \times (x) \right) \right)^\wedge 0.5 \right)$$

$$\mathrm{Total\ toll} = \$.50 + \$1.25 = \$1.75$$

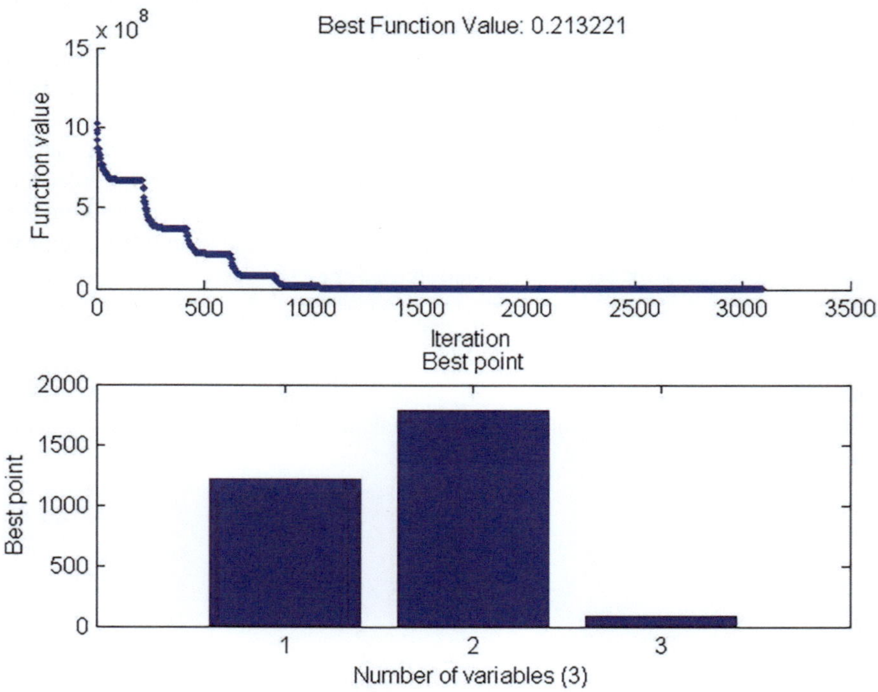

Fig. 7.1 Optimal output values, demand = 3000 veh/h

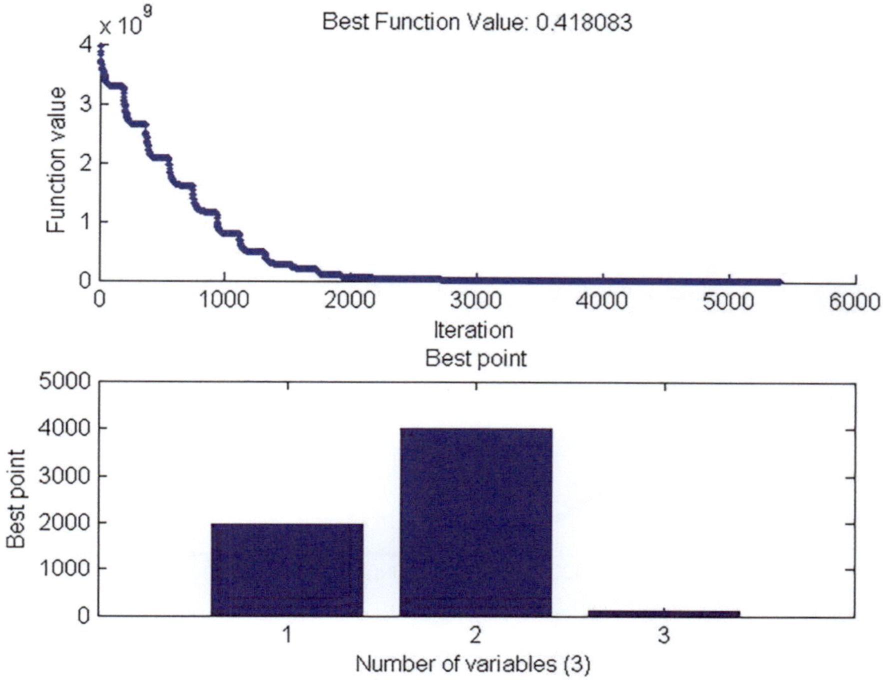

Fig. 7.2 Optimal values, demand = 6000 veh/h

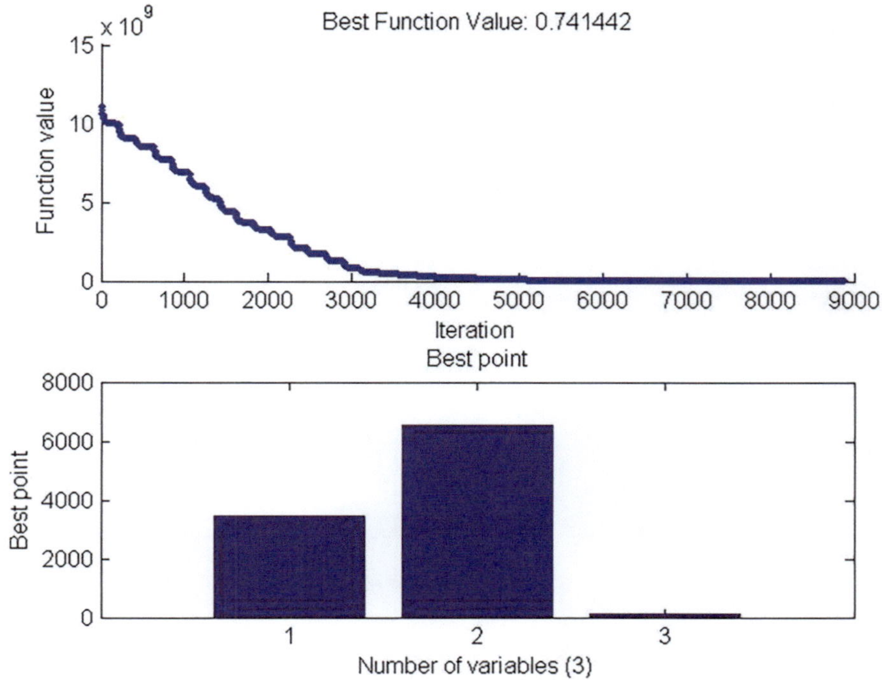

Fig. 7.3 Optimal output values, demand = 10,000 veh/h

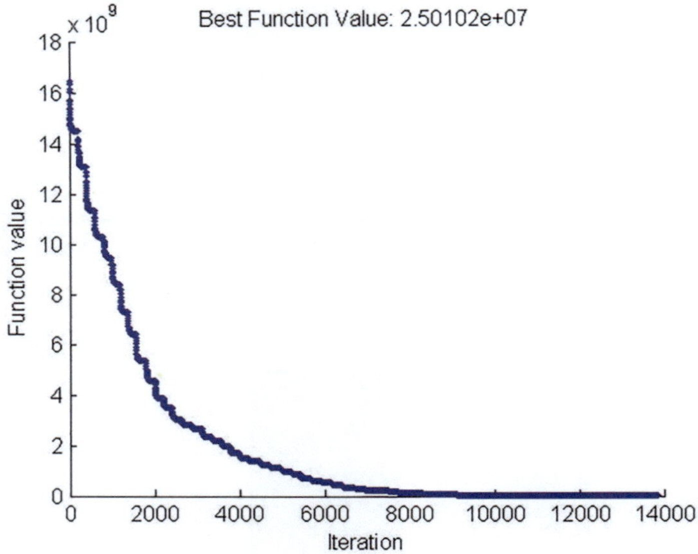

Fig. 7.4 Optimal output values, demand = 12,000 vehicles/h

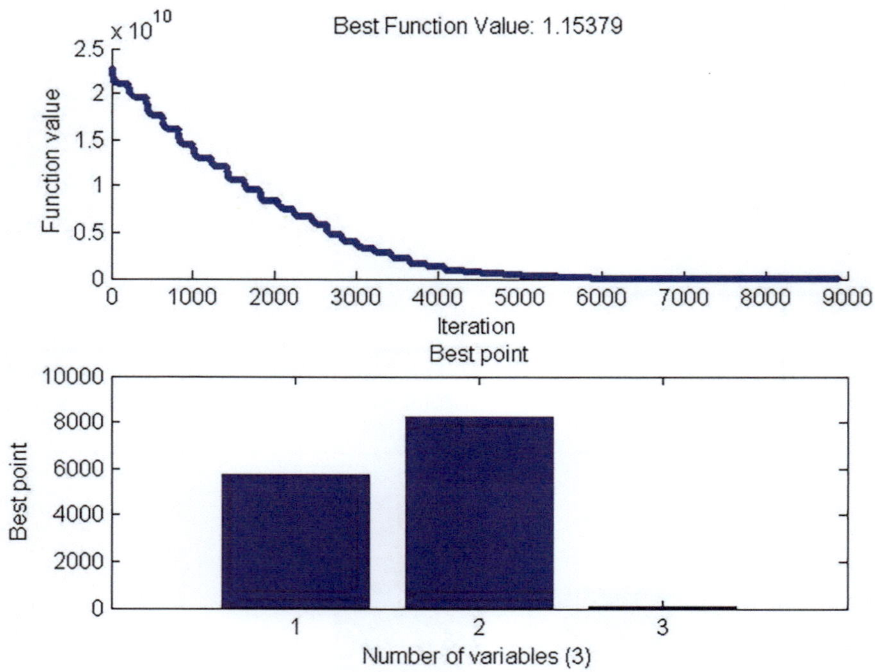

Fig. 7.5 Optimal output for demand = 14,000 vehicles/h

Table 7.1 Toll price demand = 3000 veh/h

Demand	Capacity (EL)	Optimal HOT flow ($x1$)	Optimal GP flow ($x2$)	Density veh/ mile	Base toll
3000	4000	1217	1782	25	$0.75

Table 7.2 Base toll, 6000 vehicles/h

Demand	Capacity (EL)	Optimal HOT flow ($x1$)	Optimal GP flow ($x2$)	Density veh/ mile	Base toll
6000	4000	1988	4012	41	$1.00

Table 7.3 Base toll, 10,000 vehicles/h

Demand	Capacity (EL)	Optimal HOT flow ($x1$)	Optimal GP flow ($x2$)	Density veh/ mile	Base toll
10,000	4000	3475	6525	75	$1.75

Table 7.4 Base toll, demand = 12,000 vehicles/h

Demand	Capacity	Optimal HOT flow ($x1$)	Optimal GP flow ($x2$)	Density veh/mile	Base toll
12,000	4000	4000	8000	87	$4.00

Table 7.5 Demand = 14,000 vehicles/h

Demand	Capacity (EL)	Optimal HOT flow ($x1$)	Optimal GP flow ($x2$)	Density veh/ mile	Base toll
14,000	4000	5770	8230	125	$7.00(max)

Genetic Algorithm Examples (Figs. 7.6, 7.7, 7.8, 7.9, 7.10 and Tables 7.6, 7.7, 7.8, 7.9, 7.10)

FDSA Examples (Figs. 7.11, 7.12, 7.13, 7.14, 7.15 and Tables 7.11, 7.12, 7.13, 7.14)

7.4 Data Analytics Tools

Data analytics tools and techniques extend beyond mathematical optimization such as linear programming and nonlinear programming methods.

Major areas of analytic tools such as supervised learning and unsupervised learning techniques can be applied to transportation and smart city strategies.

Regression modeling is a supervised learning technique which is very useful in predicting trends related to smart city customer behavior.

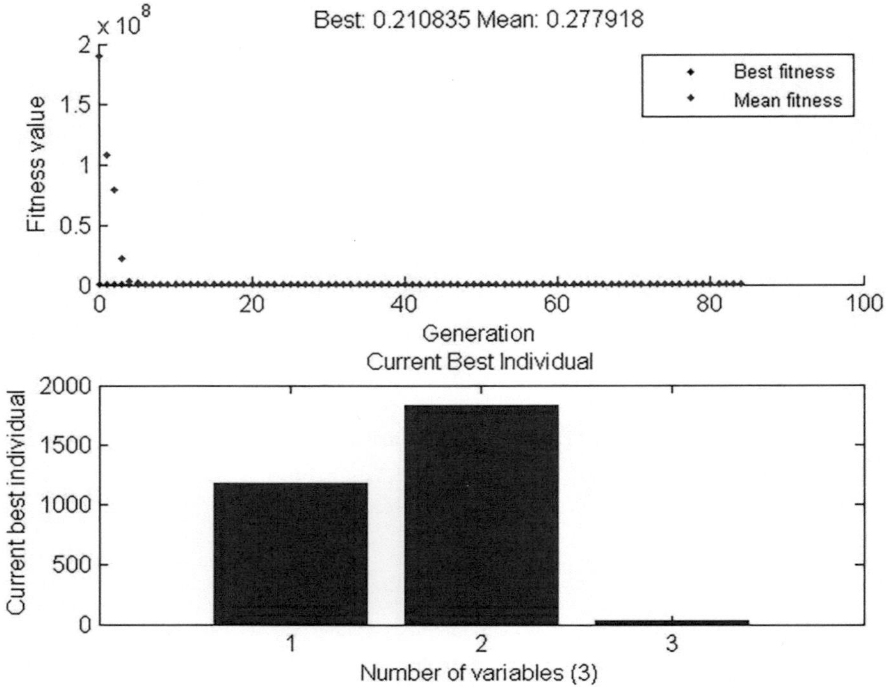

Fig. 7.6 Optimization results demand = 3000 vehicles/h

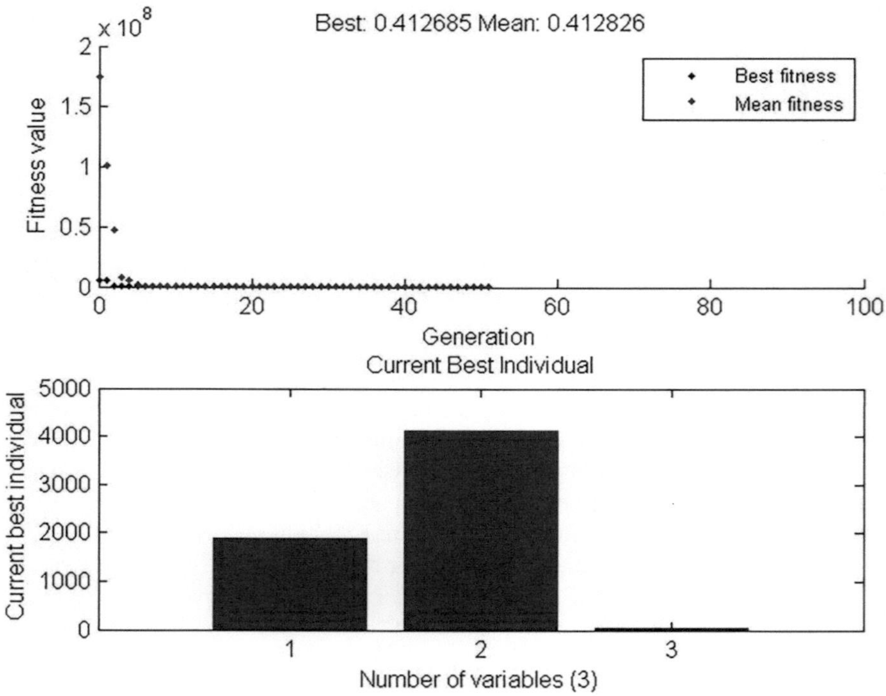

Fig. 7.7 Optimization results, demand = 6000 veh/h

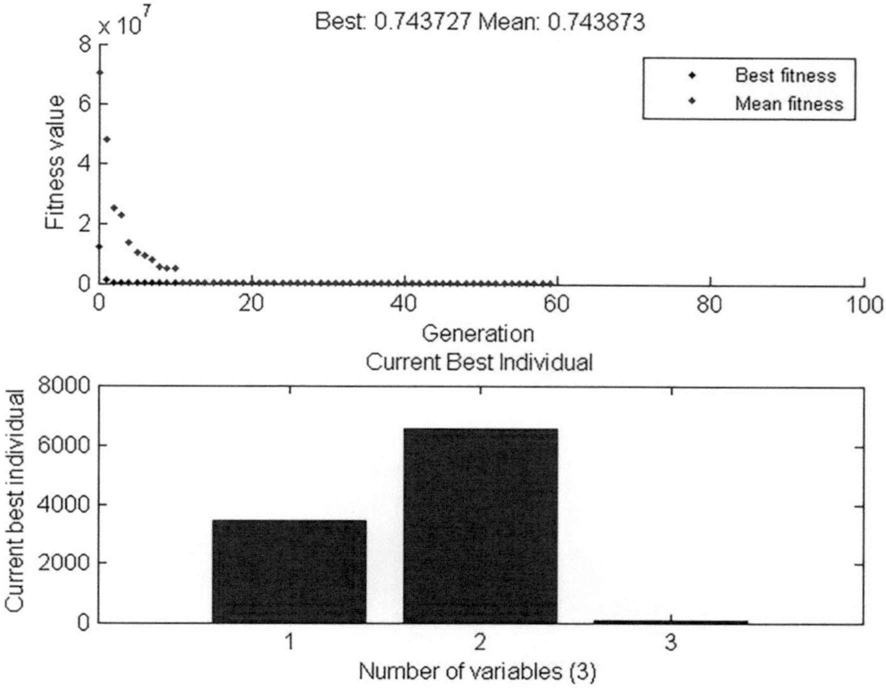

Fig. 7.8 Optimization results, demand = 10,000 veh/h

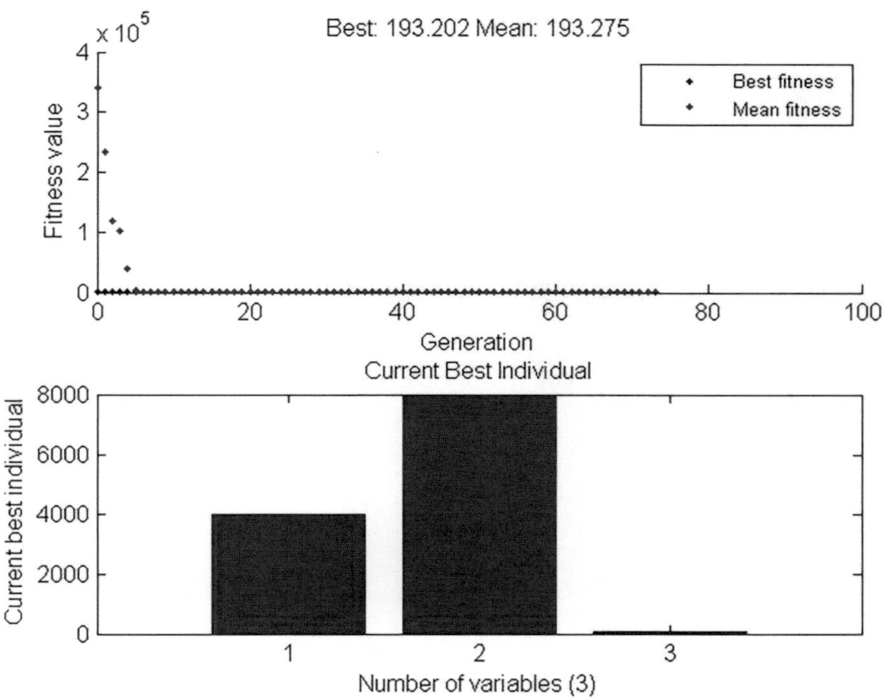

Fig. 7.9 Optimization results D = 12,000 vehicles/h

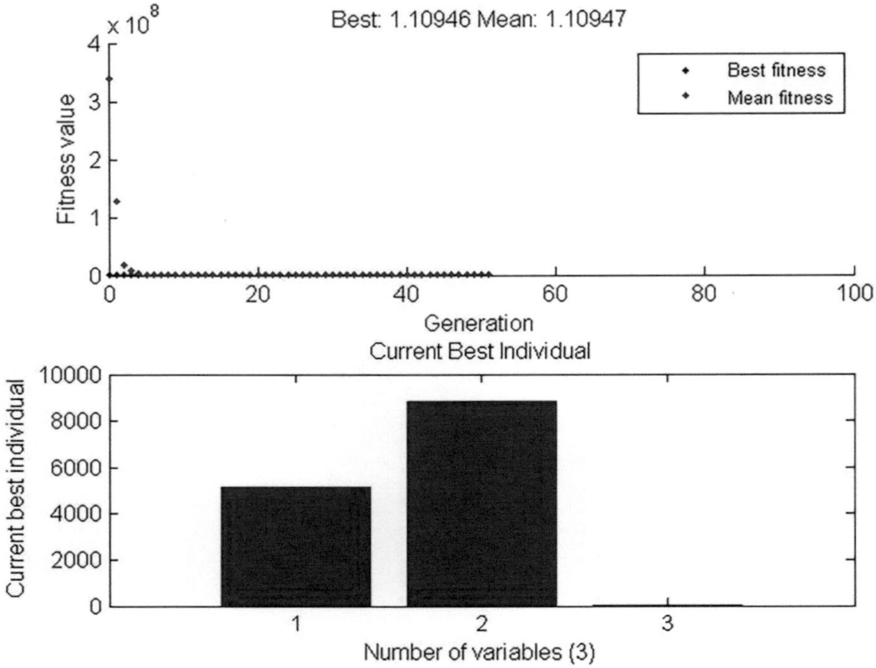

Fig. 7.10 Optimization results $D = 14{,}000$ veh/h

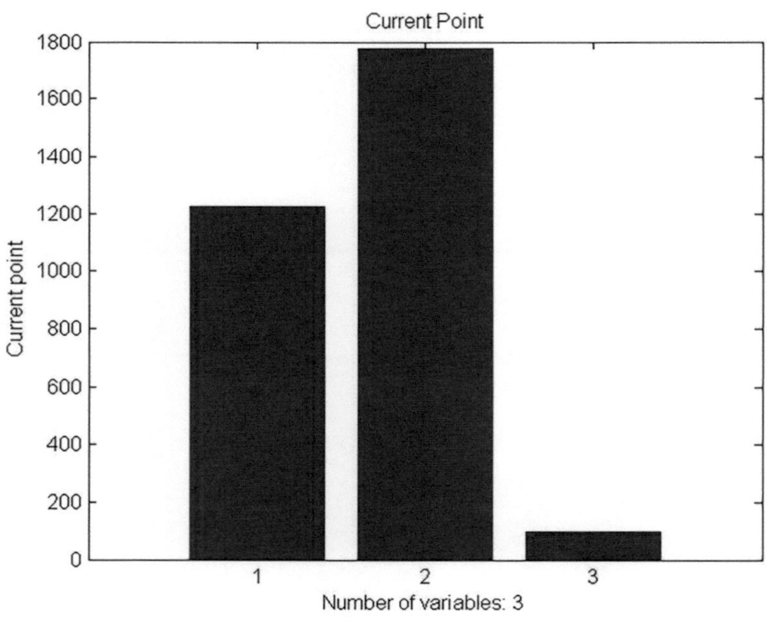

Fig. 7.11 Optimization results $D = 3000$ veh/h

Table 7.6 Demand = 3 k veh/h (genetic algorithm)

Demand	Capacity (EL)	Optimal HOT flow ($x1$)	Optimal GP flow ($x2$)	Density veh/mile	Base toll
3000	4000	1175	1825	25	\$0.75

Table 7.7 Base toll, demand = 6000 veh/h

Demand	Capacity (EL)	Optimal HOT flow ($x1$)	Optimal GP flow ($x2$)	Density veh/mile	Base toll
6000	4000	1892	4108	41	\$1.00

Table 7.8 Base toll, D = 10,000 vehicles/h

Demand	Capacity	Optimal HOT flow ($x1$)	Optimal GP flow ($x2$)	Density veh/mile	Base toll
10,000	4000	3460	6540	75	\$1.50

Table 7.9 Base toll, D = 12,000 veh/h

Demand	Capacity (EL)	Optimal HOT flow ($x1$)	Optimal GP flow ($x2$)	Density veh/mile	Base toll
12,000	4000	4000	8000	87	\$3.75

Table 7.10 Base toll, D = 14,000 veh/h

Demand	Capacity (EL)	Optimal HOT flow ($x1$)	Optimal GP flow ($x2$)	Density veh/mile	Base toll
14,000	4000	5175	8825	125	\$7.00 (max)

7.4.1 Regression Modeling

Regression modeling is a technique which uses the relationship between variables to predict the values of an outcome given the values of other related variables.

Thus, we have two sets of variables: the first is the "dependent" variables and the second "independent" variables. The independent variables are also called "explanatory" variables since they explain the reason why the dependent variable takes on a value.

Written in mathematical form:

$$Y = \beta 0 + \beta 1x1 + \beta 2x2 + \ldots\ldots\ldots + \varepsilon$$

$Y = \beta_0 + \beta_1 x_1 + \beta_2 x_2 + \ldots\ldots\ldots + \varepsilon$
Y = dependent variable

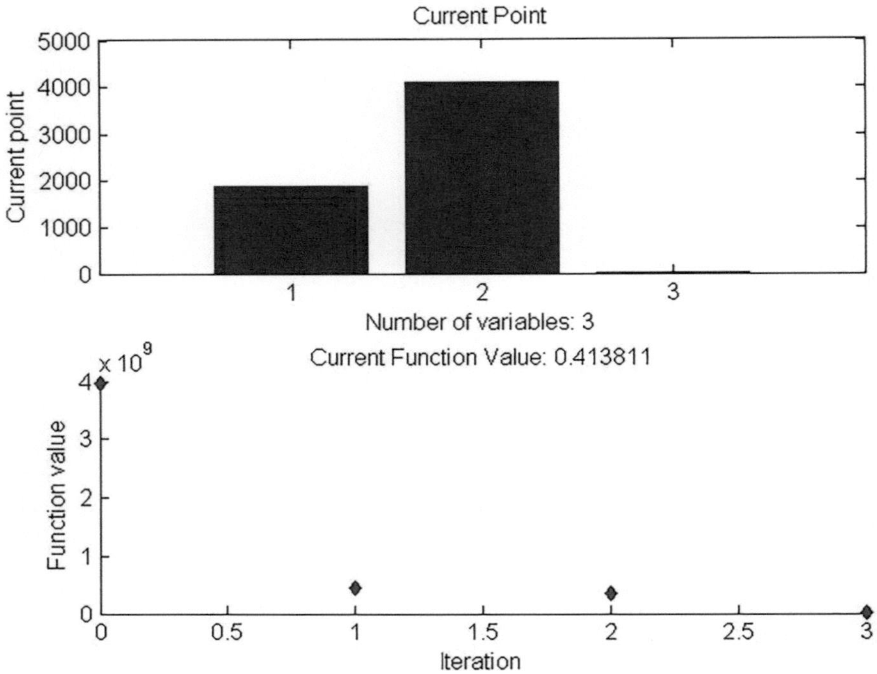

Fig. 7.12 Optimization results $D = 6000$ veh/h

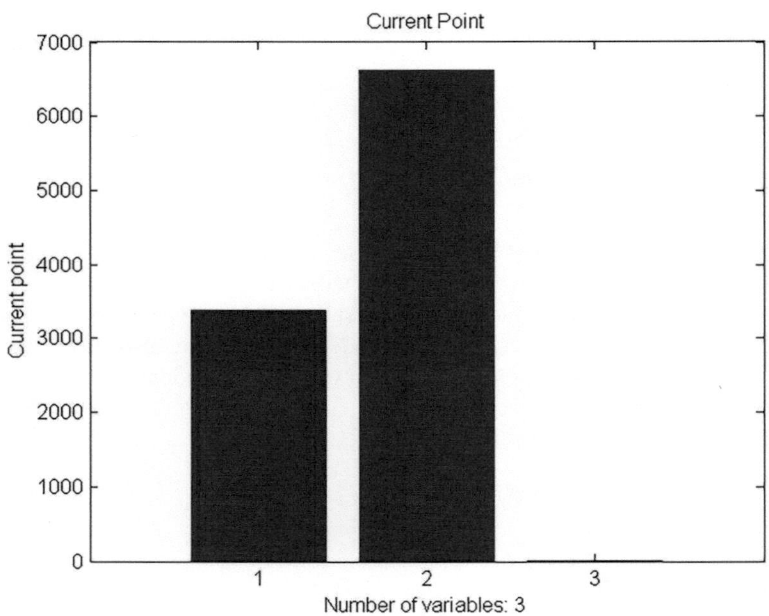

Fig. 7.13 Optimization results $D = 10{,}000$ vehicles/h

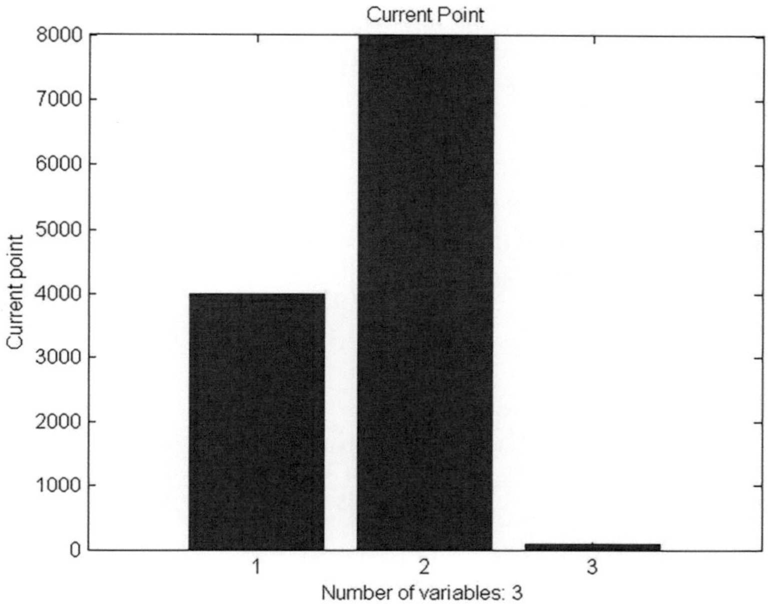

Fig. 7.14 Optimization results D = 12,000 veh/h

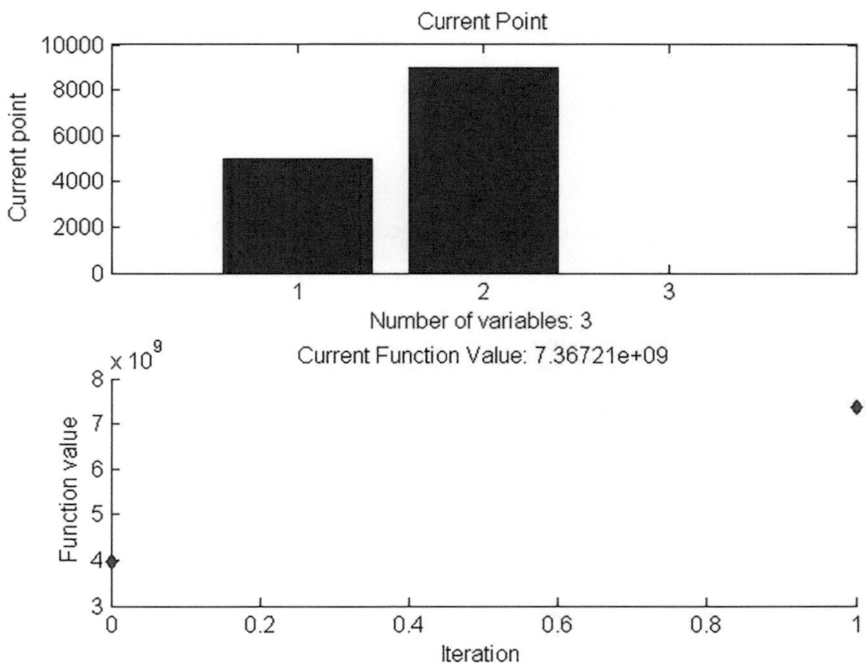

Fig. 7.15 Optimization results, D = 14,000 veh/h

Table 7.11 Base toll, $D = 3000$ vehicles/h

Demand	Capacity (EL)	Optimal HOT flow ($x1$)	Optimal GP flow ($x2$)	Density veh/mile	Base toll
3000	4000	1225	1775	25	\$0.75

Table 7.12 Base toll, $D = 6000$ veh/h

Demand	Capacity (EL)	Optimal HOT flow ($x1$)	Optimal GP flow ($x2$)	Density veh/mile	Base toll
6000	4000	1800	4200		\$1.00

Table 7.13 Base toll $D = 10{,}000$ veh/h

Demand	Capacity (EL)	Optimal HOT flow ($x1$)	Optimal GP flow ($x2$)	Density veh/mile	Base toll
10,000	4000	3375	6625	75	\$1.50

Table 7.14 Base toll, $D = 12{,}000$ veh/h

Demand	Capacity (EL)	Optimal HOT flow ($x1$)	Optimal GP flow ($x2$)	Density veh/mile	Base toll
12,000	4000	4000	8000	87	\$3.75

β_0 = constant
β_1 = a constant for variable 1
β_2 = constant for variable 2
x_1 = variable 1
x_2 = variable 2
ε = error term to include the difference between the value of Y and the sum of the explanatory variable terms (x's)

The above equation assumes there are only two factors or "explanatory" variables related to the output (dependent) variable.

So, using a regression model, we can then compute demand for travel by any mode of transit within a smart city once we know the "explanatory" variables.

Often, we don't know the explanatory variables beforehand, but we can follow a process which helps us to identify which factors are most likely related to the outcome variable we are trying to predict.

One such process begins by using historical values of the output variable and comparing them to the values of other potential explanatory variables. A correlation coefficient between the two variables (input and output) is then calculated which tells how closely the two variables are. This coefficient takes a value between −1 and 1. If the coefficient of correlation is 1, then the two variables are considered to be totally related. If the coefficient of correlation is −1, then there is an inverse relationship, and if the coefficient of correlation is zero, then the variables are considered unrelated.

Once we have a positive correlation coefficient with a high enough value, we can assume that the two variables are related and that the independent variable is a factor in explaining the values of the dependent (output) variable.

7.4.2 Simple Regression Models

A simple regression model is one where the dependent variable has just one explanatory variable. In this case, there is one coefficient constant, and the equation takes the following form:

$$Y = \beta \times X + \varepsilon$$

Where:

Y = dependent variable
X = independent variable
β = constant coefficient
ε = error term

It is assumed that the error term does not change or vary with time and is called homoscedastic. These assumptions are made for the classic linear regression model to hold.

7.4.3 Multiple Linear Regression Models

This model will utilize multiple explanatory variables X to predict the value of the output dependent variable Y.

This model takes the general form shown below:

$$Y = \beta 0 \ + \ \beta 1 X1 + \beta 2 X2 + \ldots\ldots. \beta n\, Xn + \varepsilon$$

where Y, X, β, and ε are as defined before.

Once the variable coefficients have been determined for a particular problem, the model is said to be "fitted" and can now be used to predict future values of Y given the known values of X.

The next step in the process is to test the model to determine if it meets the assumptions for the classic linear regression model.

Two tests that are typically conducted at this stage are the test for normality and the test for homoscedasticity [4]. The test for normality checks to ensure that the data follows the normal distribution for the *Y* variable and the *X* variables. When this test is done, the graph of *Y* versus *X* should be close to a straight line confirming the normality of the data.

The next test that is done is the test for homoscedasticity. This test is used to ensure that the data error values do not change over time. If the data error values changed over time, the assumption would be violated and the data would be said to be "heteroscedastic."

If the assumptions of the classic linear regression model are violated, then the next step in the process is to attempt a transformation of the variables to obtain a system which does meet the assumptions of the classic linear regression model.

One such transformation is the logarithmic transformation which can be applied to the independent variable *X* and transforms the system to a relationship between *Y* and log (*X*). This transformation is often used when heteroscedasticity is identified in the original system, and often this results in a new system which satisfies the assumptions of the classic model.

In some cases, the assumptions of homoscedasticity are not a valid assumption to make. This may be the case when there is a high level of volatility in the system such as a stock market or peak hour traffic in a transit network.

With a high level of volatility, the classic linear regression model assumes constant variance is not valid. In that case, the system will have error terms which change with time. This type of system is called a heteroscedastic model, and the test for homoscedasticity usually done with the classic linear regression model will not hold and is excluded.

Several models have been developed to handle such systems which are heteroskedastic. One such model is the GARCH (generalized autoregressive conditional heteroskedasticity). This type of regression can be shown to produce more accurate results than a classic linear regression model when the system has high levels of volatility.

For the generalized autoregressive conditional heteroskedastic model, the coefficient parameters can be estimated using advanced statistical estimation methods and fitted to the data.

7.4.4 Logistic Regression

Logistic regression is a type of regression model which can be used to predict the probability of an event occurring. Thus, the output variable is usually a number between 0 and 1. This probability depends on certain factors or independent variables which "explain" the value of the dependent output variable. Logistic regression models can help predict the probability of certain demand levels occurring in a smart city transit network.

References

1. Burris, M. (2003). The toll-price component of travel demand elasticity. *International Journal of Transport Economics/Rivista Internazionale Di Economia Dei Trasporti, 30*(1), 45–59. Retrieved from www.jstor.org/stable/42747647
2. Graham, D. (2013). *A comparative evaluation of FDSA, GA, and SA nonlinear programming algorithms and development of System-Optimal methodology for dynamic pricing on I-95 Express,* PhD dissertation. University of Central Florida.
3. Spears, S., Marlon G., Boarnet M. G., & Handy S. (2010b). *Draft policy brief on the impacts of road user pricing based on a review of the empirical literature, for Research on Impacts of Transportation and Land Use-Related Policies*. California Air Resources Board.
4. Datta, S., Graham, D. P., Sagar, N., Doody, P., Slone, R., & Hilmola, O. P. (2009). Forecasting and risk analysis in supply chain management: GARCH proof of concept. In T. Wu & J. Blackhurst (Eds.), *Managing supply chain risk and vulnerability* (pp. 187–207). Springer.
5. Soloveva, V., & Kolomatskiy, A. (2023) Valuation of unpaid time savings on toll roads using transport model. *Procedia Computer Science, 220*, 16–23. ISSN 1877-0509

Chapter 8
Practical Application of Smart City Strategies

8.1 Congestion Pricing

One of the main goals of any smart city is to reduce congestion caused by vehicular traffic as well as pedestrians. Since the early 1990s, congestion pricing has been implemented in several cities of the United States.

8.1.1 SR-91, San Diego California

One of the most widely known implementations is that of State Route 91 in San Diego, California. In this implementation State Route 91 is divided into two separate sections using flexible vertical plastic strips (pylons): one with tolled lanes and another with un-tolled lanes. The tolled lanes charge the drivers a fixed charge for the privilege of using these Express lanes. Based on the demand observed, these charges are revised periodically every 3 months [3].

8.1.2 I-95, Miami, Florida

Another example of congestion pricing occurs on the I-95 corridor in Miami-Dade County. In this phase 1 scheme, users are charged a variable fee to drive in the I-95 Express lanes between the I-395 and the Golden Glades Interchange. The goal of this project is to maintain a speed of 45mph in the Express lanes, and the fee will increase as demand increases in order to maintain this speed. Buses and high-occupancy vehicles with three or more passengers can use the Express lanes for free.

© The Author(s), under exclusive license to Springer Nature Switzerland AG 2023
D. Graham, *Integrated Electronic Payment Technologies for Smart Cities*,
https://doi.org/10.1007/978-3-031-38222-2_8

8.1.3 *Atlanta, Georgia*

Similar tolling schemes are planned or instituted on highways in Georgia. Some studies, which are ongoing in this region, include:

1. I-85 high-occupancy toll lanes. These lanes will run from Doraville to Gwinnett County Georgia. Motorcycles and emergency vehicles will be exempt from the toll.
2. I-75 high-occupancy toll lanes. The prices in this region will dynamically change based on the demand.
3. I-20 current high-occupancy vehicle lanes will become high-occupancy toll lanes.

   Prices will vary with demand.

8.1.4 *District of Columbia*

In Washington D.C., the study was completed on variable pricing along the I-270. This study examined the feasibility of using Express (toll) lanes in place of or in conjunction with the current HOV lanes.

8.2 Congestion Pricing Strategies

I-95 Express was initially developed as a managed lane facility in three phases. In the first phase, a 12-mile stretch of road from downtown Miami to the Dade County/Broward County line to the north was built with two tolled express lanes next to the median and four un-tolled general-purpose lanes to the right of the express lanes. These managed lanes existed on both the northbound and southbound sides of the highway and required use of a transponder to collect the toll. In the second phase of I-95 Express, the highway was extended north of the Broward County line to the I-595 Express (East-West) connecting commuters with the Airport in Broward County [1].

The third phase of I-95 Express will extend the roadway from Stirling Road in Broward County north to Linton Boulevard in Palm Beach County. This extension will help alleviate congestion and allow for more connection points with the light rail transit (Brightline) which operates throughout Palm Beach County.

The city of Miami is serviced by mass transit bus service as well as light rail. Rapid transit bus service is allowed on the Express Lanes between downtown Miami and Golden Glades interchange to the north.

Buses from several south Florida counties are exempt from payment on the I-95 Express managed lanes. In addition, school buses and registered vanpools as well as motorcycles are exempt from payments.

Payments on the I-95 Express can be made with a transponder such as SunPass. Other transponders which are compatible with the I-95 Express payment system are Georgia Peach Pass as well as North Carolina Quick Pass.

This interoperability allows out-of-state drivers to quickly and easily use and pay for the I-95 Express highway facility without the need to change their transponders which they purchased out of state.

In return, the states of Georgia and North Carolina allow Florida drivers who own a SunPass transponder to use their own transponder when they travel to those states without the need to purchase a new transponder (Peach Pass or Quick Pass) for those states.

The concept of a smart city is to use data combined with technology to solve the challenges in an urban environment and to create more customized solutions for different groups of travelers while improving efficiency.

With advances in technology, the growth in computing power has opened new abilities to manipulate huge amounts of data which was unimaginable only a couple decades ago. So, with this new technology, we can answer questions which we could not even attempt 20 years ago due to the lack of processing power.

The increase in data has also brought about new fields of study such as data science which includes large-scale optimization, statistical learning (supervised and unsupervised), as well as algorithms for deep learning.

Combining these theoretical tools with technologies such as the cloud and Internet of Things has brought us to a point where we can now address known problems and provide a reasonable solution within a reasonable amount of time. With this capacity, the information can be accessed, and solutions can be generated in almost real time which allow the consumer to use it to make data-driven decisions in a timely manner.

As mentioned earlier in the book, sensors have played a big role in any intelligent transportation system deployed on the nation's roadways. We now expand the use of smart sensor technology and smart cameras to show how these technologies have been applied in a smart city network including all modes of transportation.

In several smart city implementations, sensors play a critical role in data collection. Sensors have been installed on streetlights of main roads in the city as well as on traffic signals.

These sensors collect data on the number of pedestrians in a certain area, real-time counts of traffic volume per hour, as well as traffic density. When installed on traffic signals, these sensors can be networked and communicate with each other in a manner as to synchronize the flow of traffic moving between consecutive signals. By having the green phase of each signal timed in such a manner so that the traffic departing the preceding signal arrives at the next signal in time to catch the green phase of the succeeding (next) signal, resulting in reduced congestion along the roadway and a smart solution to urban traffic jams.

Smart sensors and smart cameras are also often utilized in subway systems in larger cities to count the number of commuters and provide a measure of demand at each station in the network. Since this is real-time volume data, it can be used to adjust the supply of trains at different nodes (stations) in the network to better meet the need in high-demand stations by moving trains from low-demand stations.

Thus, the system is being dynamically adjusted to achieve a system-optimal operation and results in less congestion as well as reduced travel time for the peak hour travel periods in both the morning and evening commute.

The data collected by the sensors and the closed-circuit TV (CCTV) cameras can be uploaded to the cloud, for analysis by and recommendations by human agents or in more sophisticated systems, by machine learning artificial intelligence algorithms.

The data is often open source and accessible by private firms for analytic optimization purposes.

As we have seen, smart city data can be used to generate analytic solutions to the congestion problem in urban areas. Another approach is to provide the information directly to the consumer and allow them to make their own decision to optimize their travel experience. Thus, for example, the information about a crash on a highway can be broadcast on social media networks (i.e., Twitter, Facebook live) to commuters with alternative route options so that the individual drivers can choose another route to avoid the crash scene and any time delay that route would cause. This information can also be transmitted to smart electronic message screens along the highway with time estimates for travel, to warn oncoming travelers of any delays and provide optional routes.

In providing the data to the user, the user will choose the route he/she feels is the best route for them to reach their destination in the network. This results in a "user-optimal" solution which may be good for the individual commuter but not for the system. On the other hand, when the data is processed analytically, the entire network can be assessed, and the solution to minimize congestion over all routes and nodes can be attained. This solution is considered a "system-optimal" solution since it will reduce travel times over the entire system.

The use of sensors in smart city implementations is not limited to highway congestion control or subway system mass transit but has also been applied to parking. In an urban environment, parking space is a scarce resource and can be quite costly in both terms of money and time to find a parking spot. Several smart cities utilize sensors to determine when a parking spot is occupied or empty, and the location of such empty parking spots can be broadcast over Internet-enabled apps. These apps can be accessed by commuters in the city seeking parking spots who are then able to locate and use these spaces with minimal effort saving them precious time in arriving at their destinations and parking their vehicles.

The use of sensors in smart city applications goes beyond transportation management. Sensors are also used to monitor air quality and pollution due to vehicle and industrial emissions. In some city locations where there is a high level of smog, sensors have been placed near industrial districts to measure and analyze the number of chemical pollutants which are being emitted.

These measurements are taken to laboratories for further investigation to determine the type and ultimately the source of the chemical. From these results, new emission standards can be developed and code enforced to lower pollution in the affected areas.

In another example, sensors have been used to detect criminal activity in urban areas. Some smart cities have placed audio sensors (microphones) which are web enabled at strategic locations in high-crime areas. When a gunshot is fired, the sensors immediately report the location of the gunshot and display it on a map on the website or mobile application. The police are then notified by the monitoring security firm and the shots fired event is investigated.

V2V and V2I are technologies which are major sources of communicating information of live traffic conditions in a smart city. Such technology utilizes a dedicated frequency range (dedicated short-range communication) on the radio wave bandwidth for transmission of signals. These signals are then sent to the receivers on the roadway or directly to other equipped motorists, who can then act on that information based on the information detected.

V2V and V2I communication can be used at traffic signals where the commuter is informed by the infrastructure (traffic signal) how long the red-light phase will be, allowing the driver to adjust his/her route based on the length of this timing. V2V communications can be used when there are disabled vehicles along a roadway, to warn approaching drivers that a vehicle is blocking the path and suggesting alternate paths to the destination.

V2V technology is also available for pedestrians who can be connected to the network. Smartphones in the possession of a pedestrian can connect to the V2V network and be used to warn pedestrians of fast-moving oncoming traffic, allowing them to get out of the way before an accident. This system will also warn the driver of a vehicle if an unwitting pedestrian wanders into the road, allowing the driver to stop the car prior to colliding with the pedestrian (or bicycle).

Several web applications such as Google Maps also provide near real-time reports on traffic conditions by displaying the roadway in the smart city and using colors to indicate the traffic conditions on each roadway/highway. For instance, a section of road colored red may indicate that there is congestion due to an accident or a green-colored road would indicate free-flow traffic.

With the proliferation of electric vehicles, smart cities are promoting their use as a strategy to reduce air pollution by providing incentives for owning and driving them. Smart cities provide free use of Express lanes to electric vehicles by making them exempt from tolls along priced roads. Electric vehicles also reduce the city's dependence on fossil fuels which degrade the ozone layer while also reducing noise pollution on the roadway.

The use of ride-sharing systems such as Uber, Lyft, and Zipcar has been another strategy to reduce the number of vehicles on the road in smart cities, as these car-sharing apps allow for the movement of a large number of people without the need for these people to own a car.

The growth of smart cities has been based partially on the grid with a network of sensors. These sensors can communicate with each other and with Internet technologies such as the cloud. This allows for the sharing of information with the public and analysis of the real-time data by firms involved in the public-private partnership.

Another major component of the smart city concept is the electrification of transportation. It is widely expected that within the next 30 years, the use of electric vehicles will surpass traditional gas-consuming vehicles on the nation's roadways. With widespread usage of electric vehicles for personal use as well as mass transit, the smart city will also be the green city resulting in great reduction in pollution from the internal combustion engine.

Self-driving vehicles have the potential to greatly reduce congestion and accidents in a smart city. Since self-driving vehicles can be arranged in a platoon, they increase the capacity of the road network without any physical expansion of the road or without the expense.

Self-driving vehicles therefore have the potential to improve efficiency while improving safety. When combined with electric vehicle technology, the smart city will be able to achieve some of its main goals of minimizing transportation travel time while utilizing clean energy.

Several smart cities have begun the process of implementing a plan to convert all their city vehicles from gas to electric while also developing self-driving light rail for public use at airports as well as in downtown areas as part of the mass transit system.

The US Department of Transportation (DOT) in 2015 proposed a "smart city challenge" encouraging cities in the United States to develop a plan to improve transportation efficiency while reducing congestion. The city which won the challenge would obtain a grant to implement their plan from DOT.

Seventy-eight cities in the United States applied for the challenge, submitting plans and joining the competition. The original list of cities which included Austin, Texas, San Diego, Denver, Atlanta, Orlando, Portland, as well as several others was then narrowed to seven and eventually Columbus, Ohio, was named as the winner [2].

The Department of Transportation was seeking a mid-sized US city which was large enough to be used as an example of how to address big city transportation problems in an urban environment while small enough so that the $50 million grant could make a difference in its transportation operation [2].

International locations which have developed smart cities include Vienna, Austria; Singapore; and Toronto, Canada. It is important to note that each smart city implementation is different depending on the needs of that location and the available resources.

The use of artificial intelligence (AI) to process the data gleaned from the sensors in a smart city can come with its own pros and cons. The data can be vital for the improvement of services for different groups of consumers in the smart city. On the other hand, artificial intelligence can bring on privacy policy issues if used for applications such as facial recognition and tracking of individuals without their knowledge. We need to clearly define and secure the data being collected in a smart city so that it is used for productive purposes and not to violate any individual's privacy. It also should not be used for purposes which promote inequity among user groups.

Security of the data is paramount so that unscrupulous individuals cannot access the data for the purpose of doing harm to the general public.

For the smart city challenge, the US Department of Transportation (www.transportation.gov) proposed several guiding principles for how cities would be judged. These included evaluation of how cities would [2]:

1. Develop communications network
2. Deploy and utilize sensors
3. Utilize connected vehicles

The challenge also provided several elements for cities to address in their proposal.

Some of these elements include:

(i) Urban automation
(ii) Car/sharing
(iii) Connected vehicles
(iv) Intelligent infrastructure
(v) Sensors
(vi) Usage of data
(vii) Urban analytics
(viii) Urban logistics/delivery of goods
(ix) Strategic business models
(x) Smart grid

Several cities which applied to the smart city challenge were allowed to pitch their vision for how they would transform their city if they won the grant. Some common themes included autonomous mass transit buses which were powered by electricity or natural gas. Bus stops with electric charging stations and wireless fidelity (Wi-Fi) access were also proposed.

Payment systems in the smart city proposals included mobile applications which allowed one integrated payment for several modes of transportation. Thus, commuters can pay one time on the app and use the mass transit bus and subway on the same day using the same online payment.

This ease of use would allow higher efficiency, better service, and more accurate accountability when reconciling accounts at the bank.

Partnerships with private organizations are also a key feature of any smart city development. Private companies can provide data collection and analytic services. For instance, Mobileye (an Intel company) already provides smart cameras and artificial Intelligence for bus fleets in several cities. This system monitors the traffic ahead and warns the driver of upcoming traffic problems using visual or audio alarms on the dashboard of the vehicle. These warning signals include events such as obstacles in the roadway, lane departure warnings, blind spot warning, as well as sudden braking of vehicles in front of the fleet vehicle.

8.3 ParkMobile

ParkMobile is a mobile phone app which many smart cities are using to revolutionize their parking system. ParkMobile is a great alternative to using coins to pay for parking and benefits the consumer in several ways.

The ParkMobile allows the user to pay online with a credit card for the length of time they want to park at any ParkMobile parking site (www.parkmobile.io).

Using the ParkMobile app, drivers can locate available parking spots in the vicinity of where they want to park. Once located the app allows the driver to register the parking by scanning a code using the phone's code reader. The user can not only pay for the spot but also extend the time at the parking spot without being at that location. This is very convenient since drivers who are inside a building several blocks away can perform the transaction of putting more time on the parking spot without having to physically leave a building to come on the street to place coins in a parking meter (Ramdatt, C. Personal communication, September 9, 2019) (Figs. 8.1 and 8.2).

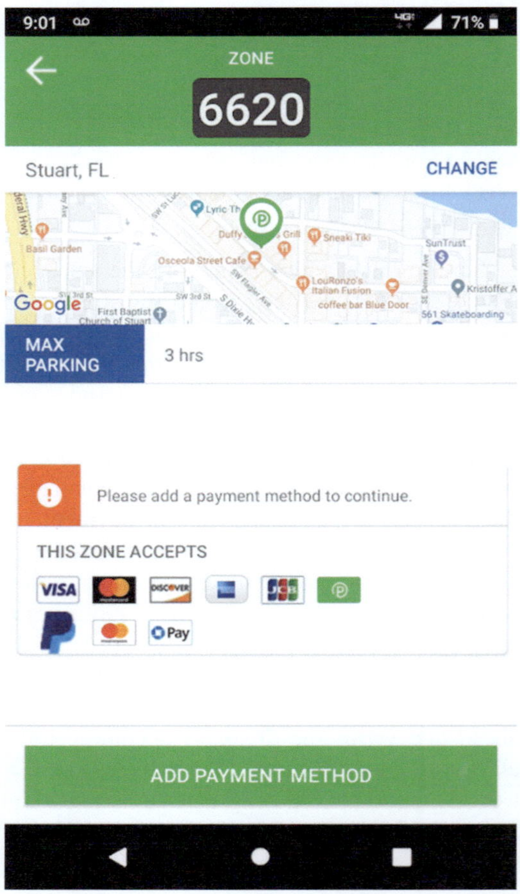

Fig. 8.1 ParkMobile app interface

Fig. 8.2 A ParkMobile sign in a parking lot in Florida

Because of the ease and convenience of using an online app for the payment transaction, many more drivers are attracted to areas with ParkMobile parking. This increases revenue and utilization of these parking spots dramatically and thus revenue increases also.

Smart cities such as Orlando and others have already adopted the use of apps such as ParkMobile or Parksmart to improve the efficiency of the parking experience in their downtown areas.

It is important to recognize that smart city concepts go beyond that of just smart transportation. Smart city includes improvement in areas such as:

Safety: Reduction in accidents and safety of pedestrians are of paramount importance in the smart city of today.

Smart buildings: Ability to control the function of the building remotely, e.g., solar-powered building.

Energy grid: Use of alternative fuels for power. Wind, solar, and hydroelectric power is increasingly becoming the backbone of the smart city.

Any reduction in the use of carbon-based fuels will reduce air pollution resulting in a greener, cleaner urban environment.

Smart city applications include water and electric utility systems. Smart meters which can be operated remotely to turn on and turn off utility services are being implemented in several locations across the United States. These meters make it easy for customers to access service as well as for the utility company to control and track the usage of services.

Many electric utility companies offer homeowners rebates and incentives to install solar panels at their homes to reduce their electric consumption. In some cases, these solar-powered homes add energy to the grid because they produce more power than they consume, resulting in a net gain for the utility system. Smart city challenge was targeted for mid-sized cities with a population between 200,000 and 800,000, and the top seven finalists were awarded a $100,000 stipend to finalize their proposals. The $50 M grant was then awarded to the city of Columbus, Ohio, in 2016.

8.4 International Applications

Outside of the realm of the US smart city challenge, several international cities have already implemented smart city concepts. Some of the major city locations on this list includes Vienna, Austria; London, UK; Singapore; Seoul, South Korea; and Dubai, UAE, among others.

Several smart city locations have targeted age and user groups to develop their smart city applications. For instance, the elderly population has been a focus point for some cities. These cities such as Singapore (a city-state) have utilized sensors to remotely monitor the health of elderly or disabled individuals who cannot easily visit a doctor's office.

With the use of smart cameras, and connected medical devices, these patients can be evaluated for physical conditions and checked for chronic conditions such as blood pressure spikes.

In addition, self-driving cars have been applied to elderly populations to assist with daily tasks such as shopping and routine errands which they may find tedious due to health issues. Young commuters who are not yet able to drive may also make use of autonomous vehicles to transport them to after-school activities in a safe and efficient manner.

Across Europe and Asia, the use of ride-sharing bicycles has reduced the number of cars on the road. Many citizens rely on the availability of these bikes to get around an urban environment. As opposed to taxi or subways, they provide a freedom to go where you want when you want without expenditure of fossil fuels from cars or trains. The effect is a reduction in congestion and travel times across smart cities due to the adoption of these pay per use bikes.

8.5 ShotSpotter

ShotSpotter is an acoustic system which sets up sensors across a broad area of interest in a city. ShotSpotter is used in several cities in association with the law enforcement agency.

An array of acoustic sensors is installed in the high-crime area. When a gunshot is fired, the sound is detected by several of these acoustic sensors. Through triangulation and artificial intelligence algorithms, the location of the gunshot can be narrowed down and sent to emergency responders who can then investigate the shooting [4].

Even when multiple shots are fired or automatic weapons are used, ShotSpotter can identify the several locations and report this to law enforcement.

As mentioned previously in the book, smart cities have adopted much of the internet connectivity for use in implementing the smart city network. In fact, the term "smart" in smart city refers to the fact that the system is connected by an information network which allows for data to be shared and used by different stakeholders in the system. The data is not just collected in real time but also made useful through data-driven decision-making. This model has beneficial impacts on the user such as a commuter on his/her way to work who can "see" traffic which is far ahead along the route and make an informed decision to change routes to a less congested one.

This same connected concept can be applied to the payment technology and system used for payment transactions. Along with mobile smart phone apps, and transponders which use radiofrequency ID (RFID) technology, the smart city can employ blockchain technology and cryptocurrency to expedite payment in the system.

8.6 Big Data and Smart Cities

We have previously mentioned some big data concepts in the book, so now we will expand on the techniques which can be applied to the data collected in a smart city.

8.7 Regression Models

Regression models are a type of supervised learning method which is most often used for forecasting future events or variable values. With the ability to collect real-time data from smart city transit systems such as subway ridership, bus ridership, pedestrian, and traffic volume/hour, we can use this data to predict what future values will be given the explanatory factor variable values.

A typical big data application could be intra-day subway ridership. For instance, a transit agency manager may have the number of riders at a subway station in the smart city at 2:00 PM on a given weekday, and he/she wants to predict the demand at that same station for the 5:00 PM peak period. Using past data and a well-tuned regression model, the ridership volume can be predicted within a small percentage error. With this information, the agency manager can adjust the supply at the stations throughout the system to ensure that high-volume stations have enough trains to meet demand and low-volume stations can have their supply reduced since high volume is not needed.

This balancing of supply with demand on an intra-day basis is only possible with the real-time capability of a smart city. Thus, we have implemented a system which is responsive to actual conditions and changes dynamically with varying needs.

This ability to respond dynamically will lead to less congestion and lower travel times across the board for the transit system. The example above can be extended to transit bus demand and supply throughout the smart city.

8.8 Optimization

Optimization models include linear optimization problems as well as nonlinear optimization models. The linear optimization model is developed with a linear expression and linear constraints. The nonlinear optimization model will have one or more nonlinear variables or terms along with constraint expressions. These optimization models have the capability to address problems which are goal oriented, that is, they can solve problems such as "minimize travel time" or "maximize throughput" as the goal. In a smart city setting, we often want to minimize congestion and travel time along the main arterials or highways to move as many commuters as possible in the smallest amount of time. With the large amount of data available in a smart city database, we can characterize the travel time between any two points by an algebraic equation and use this expression as the goal of an optimization problem. Thus, any route in a city can be analyzed given real-time data and optimal travel time throughout the network from point A to point B can be determined.

In the same way, the throughput volume on any roadway or subway station can be determined in vehicles per hour (or passengers per hour for a subway train station). This information can be used to program a mathematical model to maximize throughput in the network and respond in real time to the data collected from the smart city collection sensors. Another useful option for transit managers is that of maximizing revenue. This goal of maximizing revenue is dependent on demand and supply at different nodes in the network and developing a pricing strategy which captures the peak demand periods and intra-day fluctuations of the flow of traffic or commuters. Taking advantage of this big data information can lead to greater profit and better service for the consumer.

Several applications have been optimized for smart cities. One mass transit mobile application for bus riders displays the real-time location of the bus they are waiting on and estimated time of arrival to their bus station.

As mentioned earlier, law enforcement agencies use ShotSpotter to identify the location(s) and source of gunfire in a smart city. Another tool widely used is that of autonomous drones as the first response to emergency calls involving criminal activity. These drones can usually get to the location of the incident before the human first responders and provide immediate smart camera views of the surroundings near the incident as it unfolds.

Similarly, mini robotic devices are deployed in dangerous or hazardous situations to provide locations of possible peril and transmit live video images back to the responding staff.

8.9 Blockchain

Blockchain has been introduced as a secure transaction protocol for Internet usage for several years now. Typical payment systems involve a bank or central financial institution to receive and process payment transactions. With blockchain, there is no central unit but a peer-to-peer network of entities which maintain a record of and validate all transactions. If a record is changed in the ledger of one of the peer entities, the difference is detected by the other entities and that transaction is invalidated. Therefore, blockchain involves a standardized procedure for entering new transaction records such that all the entities in the network receive the same information about the transaction such as the amount and source of the payment once validated; each entity has the identical information about the transaction which provides a higher level of security than traditional systems.

The smart city provides web-enabled payment systems via application on mobile phones for transactions. This system consists of the consumer, who interacts with the transaction system for each of the service providers, and the service providers themselves. This smart city data network is ideal for the blockchain system since blockchain also operates within a network system. Using blockchain has several advantages for the transactions in the smart city. For one thing, blockchain transactions do not require a centralized third party so the transactions are much faster. In addition, in an open distributed network, security of blockchain transactions is stronger than a single third-party bank.

In the sections that follow, we will examine three implementations of smart cities in the United States and explore what strategies were used and whether these strategies were successful thus far.

8.10 San Diego California

San Diego, California, is a major city in California with a population of over 1,000,000 which implemented a smart city program over the last several years.

One of the key components of the San Diego smart city program was the use of sensors on traffic and street lights in the downtown area. With over 28,000 streetlights, they created a network by selecting upward of 4000 if these lights and installing GE smart sensors.

These sensors were equipped with smart cameras as well as air quality sensing ability. The live stream from these cameras was used to monitor parking, traffic volumes (per hour), as well as pedestrian traffic. The data from these sensors were uploaded to the cloud. The information collected from the cameras can inform drivers of the location of vacant parking spots, whether there is a traffic congestion on a road section or if there is heavy pedestrian volume at an intersection. Smart monitors can also be used in conjunction with streetlights to change light from red to green when there is light crossflow traffic. Thus, if a driver is alone at an intersection late at night, he or she does not have to wait at a red light if there is no traffic flowing in the cross-street.

Another use of the sensors is to provide infrastructure for connected vehicles to communicate information about traffic to the network and allow agency officials to adjust the timing of lights to reduce congestion. Connected vehicles also communicate with each other and can sense distances between each vehicle (headways) to avoid collisions.

Another strategy that San Diego implemented was the use of electric vehicles. Electric vehicles, autonomous vehicles, and electric bikes are the direction of the future for transportation by smart city planners. Thus, San Diego has invested in expanding the number of charging stations making them more accessible across the city.

Public-private partnerships with the universities in California have led to research into alternative fuels to replace carbon-based fuels. This research has led to innovations into biofuels such as algae and solar powered vehicles.

Another major partner is with General Electric (GE) which provides the smart sensors that are placed on the streetlights for monitoring of traffic.

Along with providing these smart sensors, General Electric has also provided LED lighting for approximately 14,000 streetlights, saving the city millions of dollars in electric costs per year.

This smart sensor network has been proposed to help guide a drone delivery system. The smart city has received approval for the initial test of such a drone-based delivery system guided by the streetlight smart network. The streetlight network can also detect potholes and upload this information about their location to the database. The transportation agency can then schedule repair of these potholes in a much more efficient way than before.

The traffic signals which are included in the network can be operated by the emergency first responders, and in the event of an emergency, they would have the ability to modify traffic lights to clear the path along the way to the location of the emergency by changing the timing of the green phase and synchronizing the traffic signals along the route.

Similar to the ShotSpotter application, the smart streetlights have the capacity to identify the location of gunshots when they are fired and notify the law enforcement agency closest to the incident. Smart cameras can also detect when and where someone pulls out a gun and notify the proper authorities if the gun is not allowed in that location.

The smart streetlights can also adjust the intensity of the LED light based on the ambient sunlight and the weather. Thus, if it is very cloudy during the day, the streetlights can detect the lower levels of sunlight and activate the LED light to supplement the sunlight.

Integrating the parking location service into the smart camera saves money by installing these sensors on the streetlight. By contrast, several city parking locators install individual sensors for each parking spot to determine the occupancy of that spot. This costs much more money than a smart sensor which can perform several functions.

8.11 Columbus, Ohio

The city of Columbus, Ohio, won the US DOT smart city challenge to design a smart city by merging big data and technology.

Columbus is a large city in Ohio, and the city has an approximate population of around 850,000 people, while the Columbus metro area has over 1.5 million people.

Columbus is the state capital of Ohio, one of the fastest growing cities in the nation, and has been the home to one of the largest (by population) universities in the nation, the Ohio State University.

Columbus has deployed multiple strategies to enable its smart city initiative. The city uses the acronym CASE to describe the four main areas of development of its smart city program (J. Fenning, personal interview, September 10, 2019).

The acronym CASE stands for connected vehicles, autonomous vehicles, shared vehicles, and electrification.

We will investigate each strategy independently next in the paragraphs below.

8.12 Connected Vehicles

We briefly mentioned connected vehicles earlier in the chapter.

The connected vehicle uses dedicated short-range communication technology along with sensors to allow vehicles to communicate with each other. This technology includes vehicle to vehicle (V2V) communication in which one vehicle can transmit information to other connected vehicles relating to its location on the roadway as well as speed and direction. Such information can be used to prevent head-on collisions, rear-end collisions, and lane-merge collisions. It also allows a vehicle to gain information about traffic which cannot be seen due to visual obstructions. For instance, if a car is directly following a large vehicle such as a bus (or 18-wheel truck) which blocks its view of the traffic ahead of the bus, and the car directly in front of the bus stops suddenly, this information can be transmitted to the vehicle behind the bus via vehicle to vehicle (V2V) communications. Thus, the vehicle behind the bus can "know" that the vehicle in front of the bus is stopping without seeing that vehicle and slow down or stop even if the bus which is ahead of it does not immediately stop.

In Columbus, the Route 33 corridor has implemented a system to allow for connected vehicles using V2V and V2I. The V2I infrastructure includes communication between vehicles and infrastructure such as traffic signals and streetlights which are specially equipped with electronic components which "talk" to each other. Note that this technology can also include humans such as pedestrians who are equipped with smart apps and can communicate with and control traffic lights. This helps immensely when individuals with disabilities need to cross an intersection.

Similarly, V2I technology allows for emergency vehicles such as ambulances or fire engines to communicate directly with traffic signals and change the lights if an emergency has occurred and they need to get through an intersection immediately.

Because of the multiple communication types, this technology is often referred to as V2X.

8.13 State Road 33 Columbus, Ohio

State Road 33 runs through Columbus from the southeast to the northwest (or vice versa depending on perspective) of Ohio. The section of SR33 roadway from Columbus to the northwest part of the state has been set up as a test bed for connected vehicles (https://www.thebetadistrict.com/us-33-smart-mobility-corridor/), and Columbus has strategic alliances with a number of automakers on the technology implementation for the DSRC (dedicated short-range communication) system along the road such as the Ohio State University. Moreover, several automakers and transportation-based firms such as Honda and National Highway Transportation Safety Administration (NHTSA) call State Road 33 home. State Road 33 is also called the "smart mobility corridor."

8.14 Autonomous Vehicles

Autonomous vehicles are one of the four pillars of the Columbus smart city program.

Columbus initiated an autonomous, electric shuttle bus service on a circuit through its downtown area. The shuttle bus is autonomous with no drivers but does have an operator onboard who provides information for the passenger and can take control if needed. Each shuttle bus can hold up to five passengers, and there are always two to three buses on the circuit which means one is available every 5–10 minutes.

Columbus plans to expand its autonomous vehicle shuttle bus program to a second round called round 2, which will create a new circuit around the Lindenwood residential district.

8.15 Shared Vehicles

The shared vehicle economy has grown significantly over the last 10–20 years. With the onset of such services as Uber and Lyft, shared rides have been a major revolution in transportation in many major cities. Particularly for elderly or disabled individuals who cannot drive themselves, it makes getting around much easier than taking a bus.

Car sharing or bike sharing is also one of the pillars of the smart city initiative in the city of Columbus, Ohio.

Columbus has partnered with shared vehicle services such as Lyft in both its car ride-sharing and bike-sharing services. Scooter and bike services offered by Byrd and Spin have also been major players in the downtown Columbus area transportation market. Ride-sharing service Lime also offers bike and scooter services.

Along with making it easier to get around, these services help significantly with the reduction in the use of carbon- based fuels and are green alternatives to fossil fuels like gasoline. They therefore serve a dual purpose in any strategy which a smart city implements.

8.16 Electrification

Electrification is the fourth pillar in the CASE acronym which Columbus adopted as a smart city strategy. This is a major part of its plan and was awarded a separate grant (besides the USDOT $40 M grant) from the Allen foundation to apply to electrification projects (Fig. 8.3).

The electrification aspect of the strategy includes:

(a) Electrification: increase electric scooters by 500%
(b) Goal of making electric vehicles (EV): 2% of all new car sales by 2020
(c) Educating residents through social media ads on EV usage, car dealership education, offering EV test drives
(d) Setting up EV acceleration partners for the purpose of electrification of their private fleet and transitioning to EV

Fig. 8.3 Electric vehicle (EV) charging station in Florida

(e) Some of the acceleration private partners include Ohio State University, L brands, AEP, Nationwide Insurance, and Cardinal Health. There are 65 private acceleration partners using mobility ambassadors.
(f) Private partners adopt rules on how employees commute to work sites by using electric vehicles or shared rides, reducing SOV (single-occupancy vehicle) trips through incentives.
(g) Increase city vehicles from the current EV 90 vehicles to a goal of 250 public EV by March 2020.

8.17 Private Partnerships

The smart city program has been built largely on the public-private partnerships which Columbus established with numerous companies in industry. As mentioned before, there are upward of 60 private firms which are directly involved in the Columbus program.

Some of these private companies are categorized as execution partners, while others are more of adopters, for example, they participate by incentivizing their employees to use the smart city-enabled systems. For instance, one acceleration partner company provided financial incentive to employees who changed their mode of transportation from driving their own private car to work and used the Columbus smart city electric buses to travel to and from work instead.

Other partners are considered execution partners since they help with the technology implementation involved in the smart city. Some examples of these execution partners are Autodesk, ATT, DC Solar, and Bosch which provide useful technology for the implementation of the smart city.

Bosch, which is one of the smart city partners, provides video analytics for the Columbus smart city "Experience" center. This system can monitor the pedestrian traffic and vehicle traffic outside the center in live real-time mode.

Having won the smart city challenge, Columbus has partnered with several private and public institutions to provide better service through transportation and other government agencies for its residents.

One of the private partnerships has been with NXP, to provide seamless digital payment systems for bus and mass transit in general (www.smart.columbus.gov).

NXP uses near-field communications (NFC) technology to allow commuters to use their smart devices such as cell phone or wearable digital device (i.e., smart watch) to pay the fare for the transportation services.

Near-field communications (NFC) allow for contactless transfer of information from payer to payee simply by holding a digital device (i.e., cellphone) near to the receiver (i.e., credit/debit card reader) and the user's account information can be accessed and the transaction completed.

NXP has in turn partnered with MIFARE (www.mifare.net) to provide all the components of the digital transaction processing system.

As previously mentioned in an earlier chapter, radiofrequency ID (RFID) technology has been used in transponders for highway toll collection. In addition, it is widely used in managed lanes for congestion pricing in states such as Florida (SunPass), Georgia (Peach Pass), and North Carolina (E-ZPass).

In mass transit, RFID technology has also been applied for payment at stations and bus stations.

This technology reduces the need for a manned toll plaza since the transponder transmits the account information over the radio wave and the entire transaction takes place without the commuter (or driver) stopping. This greatly increases the flow of traffic on the highway, as well as riders on the local roads.

8.18 Pivot Trip Planner

Columbus smart city initiatives include an application called Pivot for planning multimodal trips in the Central Ohio region. Pivot allows the commuter to use the application to input a destination, and the app will then plan a route with multimodal transportation. The application seamlessly integrates bus, car share, and all available types of transportation to get the user from point of origin to his/her destination.

Below is the Pivot application interface (Fig. 8.4).

Fig. 8.4 Pivot app mobile interface. (Source: pivot.io)

8.19 Denver

The city of Denver was one of the finalists for the US Department of Transportation smart city challenge. Denver is known as the mile-high city because its elevation is one mile above sea level. Denver's location makes it conducive to certain types of alternative energy sources which can be harnessed and used in its smart city program.

Solar power is one of the components in Denver's strategy to replace fossil fuels. Denver has developed solar energy farms to supplement its supply of energy for electric and autonomous vehicles.

In implementing its smart city plan, Denver has partnered with private companies such as Panasonic.

8.20 Private Partner: Panasonic

Panasonic has built an entire community of several hundred homes and families in Fujisawa, Japan, based on smart city concepts involving smart sensors, autonomous vehicles, clean energy for electric vehicles, and connected vehicle technology. The use of car sharing is a strategy to reduce congestion in downtown.

Denver has also partnered with Panasonic to use smart streetlights and solar power to develop a "microgrid" of energy for its smart city initiative. With a network of smart nodes, real-time data can be transmitted about demand for transportation to city planners and agency managers who can adjust the supply so that peak demand in a smart city is met.

Many questions have been asked concerning the use of the data collected on private citizens by the smart city sensors. Issues surrounding privacy and how the data could be used to illegally intrude on the lives of the citizens' right to privacy have surfaced across the nation. It is important to note that in the collection of data for smart cities, only the metadata is accessed. That is to say, only counts and numbers of individual cars and pedestrians are saved in the smart city database.

Information which identifies individuals or tracks their location is generally not included in the smart city database. Only aggregated information such as averages, actual amounts, and maximum or minimum values are stored for retrieval.

Smart city research includes a wide array of technology and platforms, from digital mapping software which became available in the early 1990s to futuristic applications such as the hyperloop; any system which impacts the population mass in a city to make it easier and more convenient to move from one point in the city to another is a smart city innovation. These innovations include transportation of freight as well as people.

Smart city innovations will encompass logistics, for movement of goods (i.e., FedEx, USPS, UPS, etc.), as well as public works applications such as smart garbage cans which tell the city agency when it is full and ready for pickup to usage of drones for delivery of packages.

All these innovations have a beneficial improvement on the lives of the citizens in the smart city. In reviewing smart city implementations from Columbus, Ohio; Austin, Texas; and others from the US DOT smart city challenge, some common themes have arisen. These common themes can be compiled into a list of “lessons learned” on the path to success.

Some of these lessons include:

(a) Strong leadership
(b) Partnerships
(c) Communications

Strong leaders must be at the helm for any smart city project to work. Leaders must be willing and able to make forward-looking decisions sometimes involving new, radical technology. Strong partnerships with private firms must exist to implement the technology.

Moreover, communications between all stakeholders must be vibrant and ongoing for the smart city to evolve into success.

References

1. Graham, D. (2013). *A comparative evaluation of FDSA, GA, and SA nonlinear programming algorithms and development of System-Optimal methodology for dynamic pricing on I-95 Express* (PhD dissertation). University of Central Florida.
2. US Department of Transportation. Retrieved from www.transportation.gov
3. Based on personal communication with San Diego smart city representative.
4. Retrieved from the world wide web. www.soundthinking.com

Chapter 9
Blockchain and Digital Currency

9.1 What Is Digital Currency?

Digital currency is a form of money, which first emerged in the form of credit or debit cards stored on an electronic device or as balances held on a website. Now with the advent of Bitcoin and blockchain technology, digital currencies have taken on new meaning as a medium for value exchange across technology platforms and open markets. Digital currencies come in different forms, with cryptocurrency and virtual currency being the most popular. Cryptocurrency involves the decentralized control of a network, while the virtual currency is used as a method to save and perform transactions. The most common form of digital currency is central bank digital currency, which is inflationary and directly issued by central banks.

Digital currency is an electronic representation of value used as a method of exchange. It's not a physical currency, but its properties are dissimilar to those of traditional currencies in that it has no physical form, is not printed or minted, and enables instantaneous transactions.

Digital money comes in two varieties: centralized and decentralized. Centralized digital money involves the issuing of tokens by a single entity (e.g., a bank). These tokens can be traded to others who are in turn able to spend them with that same entity. Decentralized digital money, on the other hand, features no central issuer or controller over the supply of tokens, and all users are given equal privileges as owners of the system.

© The Author(s), under exclusive license to Springer Nature Switzerland AG 2023
D. Graham, *Integrated Electronic Payment Technologies for Smart Cities*,
https://doi.org/10.1007/978-3-031-38222-2_9

9.2 What Is Cryptocurrency?

A cryptocurrency is a type of electronic currency that can be used for financial transactions. Cryptocurrencies are classified as a subset of digital currencies. A cryptocurrency is a set of encrypted data that has been engineered to operate as a medium of exchange. Through the use of cryptography and blockchain technology, each unit of currency is unique and highly secure, ensuring accurate transaction record-keeping.

Cryptocurrency is not backed by a commodity or a government. Instead, it is created as coins and tokens through a process called mining, in which computers solve cryptographic puzzles to unlock the coins. Because the coins are not directly tied to something that has value, their worth is determined on the open market. As their value fluctuates, people are attracted to cryptocurrency because of its potential for short-term gains.

Cryptocurrencies represent a new form of digital currency that utilizes algorithms to secure transactions and control the creation of new units. Cryptocurrency uses decentralized control, as opposed to centralized banking systems, where operations are conducted by a central entity. It is not managed or issued by one entity but rather a network of computers across the globe. Cryptocurrency is not backed by any government or monitored by a single regulating body or authority. Minting a cryptocurrency would mean that it is considered centralized. The same goes for

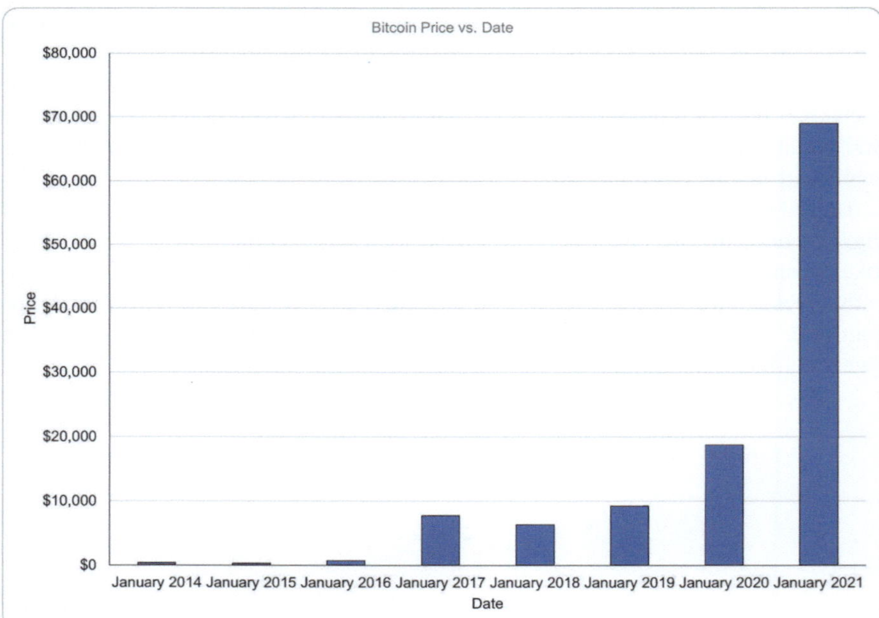

Fig. 9.1 Bitcoin price history

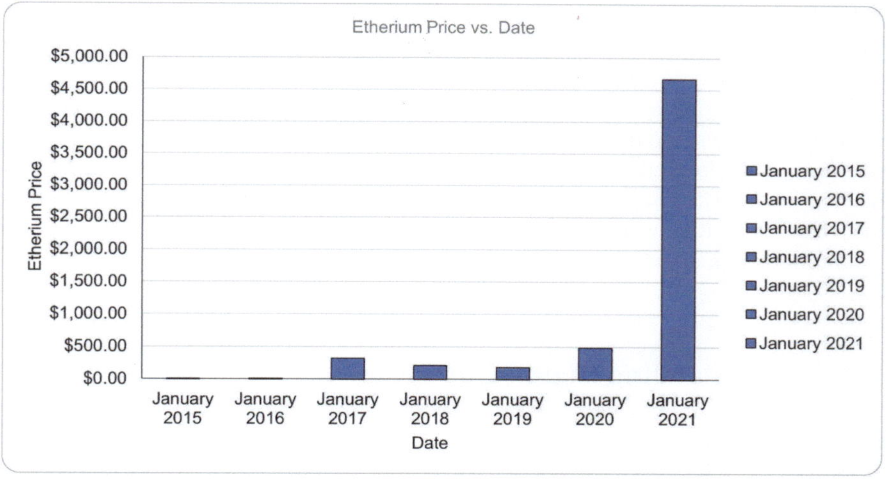

Fig. 9.2 Ethereum price history

when it is created before issuance or can only be issued by a single issuer. On the other hand, implementation of the Cryptocurrencies are digitized assets [2], used as a tradable form of money. They are issued and maintained by a decentralized network, making them impervious to the fraud and political interference common in modern industries. There are over a thousand different cryptocurrencies in circulation today—many of which can be found on well-organized exchanges with diligent security protocols. Cryptocurrency is believed to be the key to the establishment of a smart economy (Figs. 9.1 and 9.2).

9.3 Digital Currency Versus Cryptocurrency

Cryptocurrency is not all there is to digital currency. It utilizes cryptography to function. A very essential feature of cryptocurrencies is the decentralization of electronic money systems. The first and most popular of such systems is Bitcoin, based on cryptography.

David Chaum [1] introduced the digital cash idea through a research paper in 1983. In 1989, Amsterdam witnessed the implementation of the ideas in Chaum's paper in DigiCash, an electronic currency company for commercial purposes. However, 1998 saw the company filing for bankruptcy.

In the same 1998, PayPal launched service. The year before (1997), Coca-Cola launched mobile payments from vending machines. Even though the US government shut down e-gold in 2008, it had attracted several million users before then. Academia and US officials have referred to e-gold as "digital currency." Not long after the US government shut down e-gold, Bitcoin was launched (in 2009).

Attempts by the government to regulate the currency proved abortive because there was no central person or organization in control of it. The introduction of Bitcoin was the beginning of decentralized electronic currency.

The origin of digital currencies can always be traced to the 1990s Internet bubble. Digital currencies also have a history of being used for money laundering and Ponzi schemes. Such digital currencies were prosecuted for operating without MSB licenses by the US government. In China, Q coins—or QQ coins—emerged during the year 2005. Renewed interest in digital currencies has also been sparked by recent interest in cryptocurrencies.

9.4 Understanding Digital Currencies

Digital currencies are not tangible and exist only digitally. No physical form of the currency exists, meaning that it can only be held electronically.

On the other hand, dollar bills have physical attributes, such as banknotes and minted coins. Transactions involving these currencies can only take place between the people who have them.

Digital currency transactions can be executed instantly and are not subject to the same geographical limitations as traditional financial transactions. This means that, for example, a seller in San Francisco can accept payment from a buyer in Singapore without worrying about foreign exchange fees or currency conversions.

9.4.1 Characteristics of Digital Currencies

1. Digital currencies are not backed by any government or bank, and they are not physical money but rather a form of computerized payment.
2. Digital currencies can be centralized or decentralized. Centralized digital currencies can be issued by governments, such as fiat currency. Decentralized digital currencies, such as Bitcoin and Ethereum, are issued by a network of computers with no central authority.
3. The digital currency has made it possible to transfer value instantly and cheaply, regardless of location.

9.5 Types of Digital Currencies

There are several types of digital currencies, and we describe a few here.

9.5.1 *Cryptocurrencies*

Cryptocurrency payments are secure and can't be forged, but each transaction is recorded in the blockchain, the public ledger of all transactions. Due to the anonymous nature of these currencies, it can be difficult for governments to track transactions for tax purposes.

1. *Virtual Currencies*
   Virtual currencies are digital assets that can be used to make transactions with fluctuating tradable values, much like a commodity. They are tied to virtual goods, services, or other forms of value within social networks.
2. *Central Bank Digital Currencies (CBDCs)*
   These types of currencies are digital versions maintained and issued by central banks. Though they may be used like regular crypto assets in some cases, they are not necessarily decentralized. Several central banks including those in Canada, Australia, England, and Japan have considered or are considering ways to issue CBDCs; in Canada, the Bank of Montreal has started a pilot program [2].

9.6 Advantages of Digital Currencies

1. *They have fast transfer and transaction times*
   The speed at which digital currency exchanges can process transactions is a significant advantage over traditional payment networks. Because digital currencies transfer across the same network in which they are created, there is no need for intermediaries to process payments and clearing and settling transactions in a very short time frame with relatively low fees. These new methods of payment offer many benefits over traditional methods, including lower processing fees and a higher level of transparency in transactions. The use of digital currency can also make it easier to record and keep records of purchases, as well as audit information.
2. *They do not require physical manufacturing*
   Digital currencies have no physical version, so there is no need for manufacturing facilities, which makes them more cost-effective. Digital currencies are also immune to physical problems like creases, folds, or dirt that may appear in a normal dollar bill.
3. *Ease of use*
   The current monetary system enables the Fed to circulate money through a series of banks and institutions. Banks can avoid this system, however, by issuing their currency directly to citizens.
4. Cheaper Transactions
   Electronic currencies allow for smoother transactions between parties. For instance, you can buy a product from a third-party vendor without the need of going through a retailer or payment processor. Because of this transaction model,

the overall cost of making a transaction is reduced and it becomes cheaper to use such a network, as opposed to fiat currencies or other payment methods.

9.7 Disadvantages of Digital Currencies

1. *They can have storage and transaction processing issues*
   Digital currencies have different requirements from paper money. For instance, an Internet connection is necessary for any digital currency like Bitcoin because users need to access the blockchain, or public ledger, to view the balance of their accounts and complete transactions.
2. *They are susceptible to hacking*
   Digital currencies are susceptible to attacks and scams. While digital currency wallets can prevent some hacks and protect your currency, there is no guarantee that any digital currency is safe from attacks. Online wallet services can be hacked, and attackers can make changes to the underlying blockchain technology to steal funds or create a new currency that leaves you with nothing.
3. *Volatility in value*
   Digital currencies are often prone to sudden changes in price because of their decentralized nature. While this makes digital currencies appealing to investors, it also means that there is a wider range of prices for consumers to choose from when transacting. For instance, one Bitcoin was worth about $100 in 2013, but by December of that year, the same Bitcoin was worth more than $1000.

9.8 Understanding Cryptocurrency

Cryptocurrencies are digital tokens that are accepted as a form of payment. These tokens are based on cryptographic protocols, which are mathematical sets of rules used to verify the transfer of assets. The first cryptocurrency was Bitcoin, created in 2009. Over the past few years, cryptocurrencies have gained significant traction, particularly among businesses looking to save on transaction costs and individuals who want to make payments across borders without using a bank or credit card.

Cryptocurrencies can be mined or purchased from online cryptocurrency exchanges. Only a few online e-commerce sites allow customers to transact using cryptocurrencies, primarily Bitcoin, because of its popularity. However, the market value of cryptocurrencies has increased dramatically since they were introduced, and they are becoming popular among investors as potentially lucrative trading instruments.

9.8.1 *Types of Cryptocurrencies*

There are thousands of cryptocurrencies on the market today, with Bitcoin being the world's largest in terms of media coverage, market capitalization, and general popularity. While it was originally designed to function as a decentralized and peer-to-peer alternative to fiat currency, it has grown into a full-fledged remittance system, peer-to-peer payment platform, and investment strategy.

Cryptocurrencies, which are utilities that run on blockchain technology, claim a variety of functions. For instance, Ethereum's ether runs the code of decentralized applications. Ripple's XRP is used to facilitate international bank transfers.

The success of Bitcoin has led to the introduction of new cryptocurrencies. Some of these cryptocurrencies aim to become a better form of bitcoin (e.g., valuing faster transaction times or lower fees over decentralization), while others are entirely novel and use the technology for different use cases than bitcoin. These include Ethereum, Ripple, Cardano, Stellar, IOTA, and EOS.

9.9 Advantages and Disadvantages of Cryptocurrency

Tradeoffs come with every revolution, and cryptocurrencies are no different given that the revolution of financial infrastructure is at the heart of the introduction of cryptocurrencies. At present, major differences exist between what should be and what is in terms of the theoretical ideal and practical implementation of a decentralized cryptocurrency system.

9.9.1 *Advantages of Cryptocurrency*

1. Cryptocurrencies mean a new era of decentralization in the money system. Banks and monetary institutions which act as centralized "middlemen" to enforce trust and monitor transactions between two parties are no longer a necessity within this system. In this regard, the possibility of a single point of failure (like a large bank) causing a spiral of crises such as the 2008 chain reactions in the United States caused by the failure of institutions is eliminated.
2. With the decentralization of the money system in the works and the elimination of a third party such as banks or a credit card company, direct fund transfer between two parties is easier. The security of such decentralized transfers can be guaranteed by the use of transaction keys and different forms of incentive systems such as proof of work or proof of stake.
3. Cryptocurrencies like Bitcoin are used as intermediaries in the remittance economy. As more people move around the world, it becomes increasingly difficult to transfer money across borders through fiat currencies. However, by

converting a fiat currency to a cryptocurrency and then to another fiat currency at the destination, individuals can avoid the expensive conversion fees.

9.9.2 Disadvantages of Cryptocurrency

1. While digital currencies claim to be anonymous and private, they aren't. Every transaction can be traced by the agencies.
2. Cryptocurrencies are popular with criminals, who use them for money laundering and nefarious activities. The Dread Pirate Roberts case is famous, as he ran a marketplace to sell hard drugs on the Dark Web. Cryptocurrencies are also popular with hackers who use them in ransomware attacks.
3. Theoretically, cryptocurrencies should be decentralized. But in reality, they are highly concentrated. For instance, an MIT study found that 45% of Bitcoin was held by just 11,000 investors.
4. One of the biggest problems with popular cryptocurrencies such as Bitcoin is that they are mined using enormous amounts of energy. That energy consumption is a big contributor to climate change, and it also accounts for a large amount of electricity spent by miners who compete against each other to solve complex math problems. Because of the high costs involved with mining and the unpredictability of rewards, most miners, who are often multi-billion-dollar firms, pool all their resources together to ensure they have the biggest piece of the pie.
5. While blockchain technology is highly secure, companies in the crypto industry can still be hacked. Not only have numerous exchanges been hacked over the years, but even people's private keys can be stolen.
6. Cryptocurrency's price volatility cannot be avoided. Its value can fluctuate [3] considerably in the short term, and this affects the price of goods and services offered in bitcoin. When purchasing different goods and services on bitcoin, it's important to stay abreast of the value fluctuations and act accordingly.

References

1. Digicash: World's first electronic cash payment over computer networks. Retrieved from worldwide web https://chaum.com/wp-content/uploads/2022/01/05-27-94-World_s-first-electronic-cash-payment-over-computer-networks.pdf
2. Demystifying cryptocurrency and digital assets. Retrieved from world wide web https://www.pwc.com/us/en/tech-effect/emerging-tech/understanding-cryptocurrency-digital-assets.html (PriceWaterhouseCoopers)
3. Graham, D. (2013). *A comparative evaluation of FDSA, GA, and SA nonlinear programming algorithms and development of System-Optimal methodology for dynamic pricing on I-95 Express* (PhD dissertation). University of Central Florida.

Chapter 10
Implementing Digital Payments in Smart Cities

10.1 Introduction

With the intent of improving digital payments across most cities, each city can benefit from understanding different ways you can use digital payments. By creating a list of guidelines and an implementation strategy, cities can better utilize digital payments and provide better payment options for consumers by integrating digital payment methods into everyday life.

Before fully integrating payment service providers, it is imperative for smart cities to clearly articulate the following [1]:

- The strategic vision of payment's platform, mission, and goals
- Payment's use cases
- Financial modeling and payment's platform viability

10.1.1 Define Strategic Vision

A digital payment initiative requires a strategic plan that includes the following: clearly defined goals, a high-level business need, the criteria for selecting technology, the project execution plan, and promotion and sustenance after the initiative is complete.

© The Author(s), under exclusive license to Springer Nature Switzerland AG 2023
D. Graham, *Integrated Electronic Payment Technologies for Smart Cities*,
https://doi.org/10.1007/978-3-031-38222-2_10

10.1.2 Identify and Prioritize Payment Instruments and Use Cases to Be Implemented

India's digital payment landscape is radically changing for the better, with a variety of options for consumers and merchants alike. Whether it's traditional channels such as NEFT, RTGS, and cards or a more innovative option, like phone-to-phone transfers or localized offerings, brands have a lot of choices. Google, Samsung, and other first-technology companies, including telecom giants and FMCG brands, have provided the catalyst for many innovative payment solutions to emerge. These include digital wallet-based payment systems, QR code-based mobile payments, and those that can even be used without bank accounts. Another example of innovations across acquiring is the adaptation of the legacy POS system to accommodate newer modes of payments such as QR codes, UPI, and Bharat QR. These solutions offer branded payment solutions to brick-and-mortar stores that otherwise would not be able to compete with e-commerce giants. Different government agencies may have different needs since there are many systems available. Some of these options include BHIM-UPI, bank-specific UPI payment options, Bharat QR, UPI QR, and Aadhaar-based payments (biometric), viz., Aadhaar Pay, etc. An assessment needs to be completed in each smart city to see which system would be best for the target audience. Many use cases exist for digital payments in smart cities, and communities need to consider ticket sizes, transaction volumes, and the suitability of each payment instrument for a specific use case when assessing what digital payment platforms to support. A dedicated team within a city or assembled from vendors may be necessary to make these determinations, and in doing so cities can prioritize and implement digital payment platforms that provide multiple payment systems in a phased manner.

10.1.3 Financial Modeling and Viability Analysis

Financial models are essential for financing, but what about the day-to-day operations? How do businesses develop recurring incomes in today's market? The financial model will include the following elements:

- Identification of sources of investment and revenue (banks/citizens/smart city funds, etc.)
- Models of transaction pricing (slab-based pricing, flat fees, per-transaction fees, etc.)
- Business models (partnerships, profit and revenue sharing, etc.)
- The required investment for technology, infrastructure, human resources, etc.

10.1.4 Define Scope of Work for Partners

It's important to create a detailed document that clearly outlines the goals for the payment system. In addition, this document should include a "to-be" list that includes the functional and technical requirements for the payment system that is expected to be implemented. It's also crucial to identify stakeholders and their role in creating this project plan.

10.1.5 Define Governance Model for Digital Payments

A governance framework is necessary to ensure the safe implementation of any system, especially payment systems and the systems they touch. Onboarding policies are required for all new payment providers, along with other documentation, before a new provider can go live with a participating merchant. These governing frameworks and policy documents cover how disputes should be handled, how fees should be calculated and collected, along with other business rules that must be followed by all parties involved in the payments process. Appropriate systems use and control:

- Users' access and onboarding rights
- SOPs for customer onboarding, creating, managing, and closing requests
- Security of data and information

10.1.6 Selection and Onboarding of Business and Technology Partners

Vendors should select their customers fairly and transparently. The best way to accomplish this is through a request for proposal (RFP) that outlines what the company needs and how vendors can prove their product capabilities. The RFP should not be limited to a shortlist of prequalified companies but should be available to any vendor with a product that fits the scope of the project. The selection process largely depends on factors like:

- How fast banks and payment processors offer a variety of services to companies looking to facilitate quick payments. NEFT (National Electronic Fund Transfer) and RTGS (real-time gross settlement) are examples of real-time payment systems that may speed up the payment process for a business. Concurrently, businesses will have access to more customers with the help of payment processing companies that can accept credit cards or other types of payments from their websites.

- Whether the system has an authorized payment service provider who can provide the system, for example, a Payment Processing Organization Unit (PPOU) licensee who can provide stored-value solutions or a Prepaid Payment Instrument (PPI) license holder who can provide mobile wallets or smartcard-based prepaid solutions.
- Complete range of payment services offered by banks or technology providers: aside from banks that may facilitate most of the payment solutions, consideration can also be given to a few technology players specializing in particular payment types (wallet solutions, POS, doorstep services, etc.) depending on use case requirements.
- Provision of competitive pricing structure by vendors: for payment projects that demand heavy financial investment, public-private partnership (PPP) models can also be considered. For instance, the AFC system for fare collections under the National Common Mobility Card framework was set up by a few transportation authorities teaming up with banks using the PPP model.

10.1.7 Solution Development and Promotion Strategy Testing

The solution should have an agile design and consist of separate units that perform a specific function. Each unit is designed to be scalable and will require an asynchronous messaging system to route data between the modules. It should consist of separate units that are each specified for user onboarding, creation of service requests, workflow management and processing, transaction processing units, accounting units, databases, and exception handling. Before any system goes live, it should be thoroughly tested by a variety of users and stakeholders to ensure that it meets its intended purpose and that all of the necessary information has been gathered. It should also be stress-tested against certain scenarios, such as unexpected spikes in traffic or new types of usage that were unanticipated during development. All potential security risks should also be identified and remedied before a product is launched.

Monetary incentives such as cashback, tax incentives, loyalty programs, etc. but to mention a few, are effective measures that can lure citizens to use digital payment systems. This might prove especially applicable to attract and convince technology-averse citizens, those who are reluctant to pay convenience fees or might be skeptical about purchasing technology like high-end phones and personalized cards for digital payments. However, monetary incentives do not provide a complete solution to this challenge. Citizens still need to be educated as regards ease of use of digital payments, hidden costs of cash, benefits and incentives for adopting digital payments, etc. In light of this, promotional campaigns can be designed toward achieving such educational purposes, while promotions can be planned around festivals to digitize those payments.

10.1.8 Project Launch

The post-UAT phase, when a system is migrated to production, presents a unique set of challenges. To enable business users to make the most of their new software, they will need a program that walks through the processes and teaches them how to use the software. Customers should be trained on any materials provided along with the software, such as guides or manuals. These materials are critical for long-term success with a software implementation.

10.1.9 Regular Monitoring and Tracking of Digital Payments

Governments need to continuously monitor the progress and growth of transactions that happen through digital payment systems, like debit cards and credit cards, to best monitor their impact on the national economy. These dashboards will be used to give senior management a clear overview of what digital payment initiatives are taking place in the country.

Any of the following approaches can be deployed by smart cities to the onboard payment service provider [2]:

- A proposal evaluation process request
- Direct onboarding from empaneled vendors
- Partnership models (revenue/profit-sharing models)

10.1.10 Factors Mitigating Digital Payments and Way Forward

10.1.10.1 Lack of Strategic Vision

Challenges: Because of the lack of a common strategic vision, most digital initiatives do not realize their potential benefits. Employees across departments feel reluctant to work with digital platforms, and they often fail to reap the full benefits of these initiatives.

Mitigation planning: To manage digital payments, businesses need to carve out a road map that clearly defines their short- and long-term goals. Based on this road map, companies should align their digital payment initiatives with the goals that they have set up for themselves. Every member of an organization needs to be well informed about the benefits of digital payments and their implications for the company.

10.1.10.2 Sustainable Financial Model

Challenges: Digital payment initiatives are often not planned properly, which results in a significant financial impact. Many companies do not have the proper resources to make such initiatives successful, and senior management often doesn't allocate the necessary budget to make them work. Many departments lack awareness of the monetary benefits of digital payment initiatives, and they think of it as a cost center rather than an opportunity.

Mitigation planning: A sustainable model for digital payments needs to be implemented within a reasonable time frame. This can be achieved by developing an e-commerce platform that allows customers to create accounts and make online transactions with ease. Once the platform is created, it can then be used as a stepping stone to further develop the platform into a full-blown digital payment service provider (PSP). The success of digital payments largely depends on a strong financial model that is aligned with the goals of both companies and government entities. All parties should determine how to incentivize participants in the system, as well as how to generate revenue.

10.1.10.3 Reducing Reliance on Manual Processes

Challenges of end users Organizations can rely on manual processes for small transactions and low volumes of service requests, but as the number and frequency of transactions increases, so does the risk of error, which impacts the delivery of those services.

Mitigation planning: For various payment and receipt transactions, multiple digital tools are necessary to reduce the reliance on manual work. This includes electronic billing systems and reconciliation systems that ensure efficiency and accuracy. A centralized system is needed to digitize all processes so that monthly bills can be automatically generated and reconciled with the customer's receipt.

10.1.10.4 Lack of Digital Data

Challenges: Digitalization of government data is necessary for the success of digital payment initiatives. The lack of digitalization discourages people from adopting the new payment platforms.

Mitigation planning: Leading municipal corporations have utilized the services of outside vendors to digitize their existing information and convert it into digital formats. These products, along with being more user-friendly, will be accessible to future generations, facilitating the passage of time while minimizing the risk of data corruption.

10.1.10.5 Lack of Infrastructure

Challenges: The infrastructure requirements for an Internet of Things, telecommunications, and a network are currently significant. Providing these technologies to rural communities is just as challenging as providing them to urban areas.

In the absence of the latest technology such as smartphones and computer devices with accessories like POS, fingerprint machines for biometric authentication, stable/consistent mobile network, high-speed internet connectivity, etc., any digital payment initiative is short-lived.

Mitigation planning: For a city to be truly smart, it must have the necessary infrastructure and amenities in place. The government bodies work with the smart city to assess existing infrastructure and plan out what improvements need to be made. This can lead to long-term savings for both consumers and the government, which makes it in everyone's best interest that this process be transparent and efficient.

10.1.10.6 Lack of Awareness and Motivation

Challenges According to a survey by Standard & Poor's Financial Services LLC, there is confusion and apprehension when it comes to using digital payments. While 76% of adults in India are unable to understand key financial concepts such as interest rates, inflation, and currency exchange rates, they still want to make online purchases. Even though the modes of payment are digital, there is a lack of awareness regarding the ease of transactions. More so, people are unaware of digital payments in rural and semi-urban areas.

Mitigation planning: Governments, banks, and regulators should engage in a comprehensive digital literacy drive that includes marginalized and underprivileged communities to ensure that everyone is familiar with how to use the country's digital payment system. By working together, the different parties involved in the creation of a payment system will be able to reach mass markets and ensure that everyone has access to this new technology.

10.2 A Look into Examples of Smart Cities

Four cities have begun to embrace digital payments [3] even though they've all progressed at different paces and in different fashions. The reasons vary, but the consensus is that they move at a faster pace than other regions. In addition, electronic payment technologies have expanded over borders. By way of example, Tallinn, Estonia, and Helsinki, Finland, illustrate this phenomenon.

Both Estonia and Finland have been leaders in e-governance development. Information and communication technologies have been used for administrative processes prevalent throughout both countries and aid in the growth of interregional solutions. Their comparable ICT infrastructure has drawn comparisons between the two countries. Both countries use information communication technology, as well as banking on the Internet. Cross-border operations have become more appealing as the amount of interaction between borders has increased globally. The major reason for this is that similar cultures and languages, proximity, and high levels of interaction tend to make it more convenient for businesses to work with partners in other countries.

10.3 Hong Kong

Due to early adoption of smart city developments, Hong Kong has been ranked as one of the world's smartest cities. Their success with smart cards, starting with the Octopus card in 1997 and later adopted around the world, shows how quickly they embrace new technology. This trend continued with the introduction of online payments, which followed their success with offline payments by being quickly introduced across all public transportation and parking, and then into supermarkets and other nearby businesses. This payment method could be used in supermarkets and restaurants, and about 13 million transactions were processed in a day (as of 2015). Octopus is currently used by over 28 million Hong Kong residents, making it the most popular e-payment solution for Hong Kong residents. It is accepted as payment in over 6000 retail outlets and can be used to pay for both public transport and tickets to various events throughout Hong Kong. A growing number of companies also accept Octopus at their checkout systems, with over 70,000 point-of-sale devices installed across the city-state.

EPS and PPS systems, two types of electronic payments have given consumers a way to pay for anything through their cell phones (eps.com.hk). The EPS, which has over 3 million daily users, according to the company website. The EPS Company and the Hong Kong Telecommunications Holdings Limited have partnered to deliver a reliable e-payment system that allows customers to make payments over the phone or online. This saves not only time but also ensures that customers have peace of mind.

Gross settlement systems have been implemented to facilitate electronic transactions. Policymakers have prioritized financial infrastructure to make electronic payments accessible for SMEs in line with the annual budget. Hong Kong's Financial Services and Treasury Bureau has created an environment where newcomers to the industry, like mobile payment startups, are allowed to compete with traditional payment services. Regulatory bodies aid local businesses by putting in place initiatives such as e-check, electronic bill presentment and payment (EBBP), stored value facilities (SVFs), and retail payment systems (RPSs).

To make Hong Kong a successful smart city, the government must first focus on its identity as an International Financial Center. By blending technology and finance,

Hong Kong can create a seamless ecosystem between the two sectors that allows for higher connectivity and better information flow. To ensure this happens, various government agencies should work together and set up a unified brand to oversee all of Hong Kong's smart city initiatives. The Central Policy Unit (CPU) recognizes that without a streamlined and centralized approach to developing and integrating with the realm of information communications, Hong Kong is limiting its economic progress. To reduce costs and improve efficiency, CPU advocates for a separate agency (IDA, currently IMDA) to be established for information communications development and integration.

10.4 Singapore

The Monetary Authority of Singapore (mas.gov.sg) has been studying how technology can be used to drive the city-state's financial sector forward, making it a smart financial center in the process. From initiatives like Project Ubin and the implementation of Robo-advisory services, MAS is looking globally at what other central banks are doing while listening to stakeholders within the country to figure out what people want from their money. Channel News Asia reported that the financial sector dominates information and communications technology in Singapore. The Development Bank of Singapore is investing over 600 million Singapore dollars in ICT every year, compared to the 2.8 billion Singapore dollars that the government spent on ICT tenders during the fiscal year 2016. At the turn of the twenty-first century, the Infocomm Development Authority (IDA) and Media Development Authority (MDA) had begun working on a contactless smart card standard for retail, and this was eventually implemented in 2002. A consortium of local telecommunications companies worked together to develop a nationwide mobile payments platform that would allow consumers the ability to execute financial transactions via their mobile devices.

The fourth Information and Communications Technology (ICT) Roadmap released in Singapore in 2016 outlined the intention to introduce an EZ-Link card, a contactless prepaid card for public transportation that would later be used to make micropayments using near-field communication technology. Another innovation involves NETS Cash Card, an ATM card that is accepted for payment at more than 55,000 retail outlets across Singapore. Money 2.0 is a term used to describe new payment-based technologies on a nation's infrastructure. It involves the use of smart chip cards and the development of nationwide electronic and mobile payment infrastructure. The latest version of the ICT Roadmap also recommends that e-payments be developed as part of Money 2.0, which could more closely characterize the next-generation digital economy.

Electronic payments have been positioned as a key sector for financial technologies by the government of Singapore. A recent report commissioned by MAS (Monetary Authority of Singapore) details the percentage of cash and cards as well as their ratio. Almost 67% of transactions are made with cash; 59% of total transactions are made with cards. In addition to the hefty cost of ATMs and other in-person payment terminals, Singapore has a high cash circulation rate of 8.8 percent compared to 4.4 percent in Australia. Following the suggestion of KPMG, MAS released plans for how it will help regulate and govern systems for e-payments. It then opened a review on government investigations for reform in the following areas:

- Creating a payment framework would allow payment service providers to determine how to respond to consumer needs, which will be activity- and risk-based.
- Establishing a National Payment Council.

The Infocomm Media Development Authority (IMDA) is one of three authorities in Singapore that work to promote financial services through technology. IMDA and the Monetary Authority of Singapore (MAS) have forged a close relationship to protect consumers, take measures against cyber-security threats, and ensure citizens have access to the Internet. E-commerce systems are still underutilized in the region, despite some infrastructure having already been established. For instance, a system called FAST was developed in 2014 and enables real-time interbank transfers. On a more microlevel, companies are starting to employ mobile wallets to solve payment problems. The LTA recently started a new partnership with Mastercard for card payments on public transport. By enabling cardholders to pay for transportation fares with their debit or credit cards, the LTA hopes to make it easier than ever for commuters to pay their way around town. In addition, Mastercard has been working closely with banks and mobile technology partners to roll out a range of services, like cardless transactions.

10.5 Tallinn-Helsinki

Tallinn and Helsinki, the capitals of Estonia and Finland and two of the best-known European cities that are only 80 kilometers apart, have a lot in common. Approximately 1 out of every 1520 Estonians lives in Finland, and Estonians travel to Finland as often as Finns travel to Estonia. The two countries share languages: they are both Finno-Ugric [3].

Two small European countries, Estonia and Finland, have long been global pioneers in digital innovations. Each has its own unique story, but the fact stands that they are home to some of the most influential brands in recent memory—Nokia and Skype, respectively. Known as one of the most innovative e-governments in Europe (the Digital Economy and Society Index 2016), Estonia is a force to be reckoned with when it comes to digital technologies. While many countries are working to implement a secure digital identity for their citizens and connect databases securely over the Internet, Finland has already made it possible for each person in the country to have his or her digital identity. Additionally, data from government agencies can be shared between them without risk of unauthorized access via a highly secure network. While other European cities have taken steps to become smart cities, such as Helsinki, none can match Finland's progress.

At the heart of the Estonian digital economy is a secure public-private system that provides smooth and user-friendly data exchange (e-estonia.com) between citizens and private companies via the use of electronic identity, which is used by 93% of Estonians. In addition to the thousands of public databases, all large-scale government services are linked together (via secure transport) using an open-standards protocol called X-Road. The X-Road allows the city to consolidate a wide range of municipal services in one location, making urban service delivery more efficient and transparent.

References

1. *Smart cities: Digital solutions for a more livable future*. McKinsey & Company.
2. *The Digital Transformation - of Smart Cities and Smart Communities*. Microsoft.
3. Soe, R.-M., Mikheeva, O., & Tallinn University of Technology. (2017, May). Combined model of smart cities and electronic payments. In *2017 Conference for E-Democracy and Open Government (CeDEM)*. https://doi.org/10.1109/CeDEM.2017.11

Chapter 11
National and International Smart City App lications

11.1 Introduction

Earlier in the book (Chap. 8), we examined smart city implementations in a few domestic locations. In this chapter we will visit several other locations in the United States and internationally.

In recent years, there has been a surge in many cities adopting smart city technology and internationally several notable applications.

Singapore, Toronto, and Dubai are all international locations where city managers have taken active steps to modernize their traveler experience with smart city innovations and implementations.

In the United States, the Department of Transportation initiated a "smart city challenge" [1] in 2016. Over 70 cities submitted proposals to be selected as the winner of that challenge.

Of those cities, many have implemented real smart city technology although they did not win that challenge.

It is very interesting to notice the path that each city took to create their smart city. Each city evaluated their customer needs and developed their own strategy to meet those needs with smart city innovations.

Some common approaches which cities take to implement their smart city is to focus on some particular technology (or group of technologies) and hone their skills at implementing that technology across the city.

Typical technologies include Iot (Internet of Things) which a smart city may use as a source of data, another is electric vehicle (EV) technology, and another is sensor technology.

Smart city managers also look at data collection methods and assess how the data can be collected from the infrastructure and network setup for the public. Questions which must be answered include the following: How will the data be stored? What type of data will be collected? Who will have access to it?

© The Author(s), under exclusive license to Springer Nature Switzerland AG 2023

D. Graham, *Integrated Electronic Payment Technologies for Smart Cities*,
https://doi.org/10.1007/978-3-031-38222-2_11

11.2 Kansas City

Kansas City was one of the seven finalists in the US DOT's "smart city challenge." Kansas City embarked on several initiatives to implement smart technology into their city infrastructure. Following their strategy of electrification, Kansas City built an electric streetcar system.

The streetcar uses overhead electric cables for power. It runs along a route approximately 2 miles long and was built at a cost of over $100 M. The streetcar is expected to relieve congestion in the downtown center of Kansas City while reducing pollution through carbon emissions.

One of the highlights of the streetcar project are the digital information kiosks which line the route and are located at stopping points of the streetcar.

The digital kiosks are information sources not only about the streetcar but also about the downtown and local businesses in the area. The kiosks are touchscreen displays which provide information about local restaurants, sports, weather, and general downtown Kansas City business information. The information is called "hyperlocal" since it covers the areas which are directly in the vicinity, usually within a few blocks of the kiosk.

Along with these initiatives, Kansas City has partnered with private organizations such as Cisco and others to develop a sensor network for data collection of local traffic information and to provide Wi-Fi service for certain communities.

11.3 Austin, Texas

Austin, Texas, was listed as one of the seven finalists of the US DOT smart city challenge (www.austintexas.gov). Austin, the capital of Texas, is home to the University of Texas at Austin and several high-technology companies.

Austin has deployed an Internet of Things (IOT) strategy for its smart city network.

A large number of streetlights have been converted to smart streetlights by addition of sensors.

The sensors can provide information on traffic flow, pedestrian count at intersections, and parking spaces.

Smart sensors can also be integrated with connected vehicle technologies. Thus, cars traveling along the roadway can transmit information such as speed, acceleration, and direction on the sensor network which is uploaded to a central database.

This additional information means that not only do we have real-time information about the traffic count, but also speed of vehicles, braking, and change of direction are also known instantaneously.

Austin Texas has implemented a smart energy grid system as part of its smart city initiative. New projects have been deployed with neighborhood participants who have solar panels installed on their property. These solar panels produce electricity for the homes, and any excess energy is supplied to the public utility grid.

The solar production site by constant communication with the grid provides information on the actual energy usage at each location. Thus, if a particular location is underutilizing power from the grid during peak hour on a particular day and time, then the utility can shift the load to another home which is using more power to meet that demand with the excess supply from underutilizing homes.

In addition, an expansion of electric vehicle technology and charging sites have been a strategy in the Austin smart city implementation plan.

11.4 Atlanta, Georgia

Atlanta is a major hub for transportation in the United States. The Atlanta-Hartsfield airport has been named the busiest airport in the United States several times by transporting the most people per year through its terminal. The location of the city puts it within a few hours reach of several major destinations along the eastern United States. Therefore, it is an ideal place for logistics and freight to call home.

Atlanta is also the capital of the state of Georgia and home to the Georgia Institute of Technology.

Through partnerships with Georgia Tech, Atlanta has deployed the North Avenue Smart Corridor (atldot.atlanta.gov).

The North Avenue Corridor runs next to the college and is a testbed for several smart technologies.

Autonomous vehicles have already been deployed along the corridor, which has installed infrastructure technology capable of guiding vehicles along the route. A network of smart cameras is located at intersections along the roadway. These smart cameras transmit live video to the database for analysis. The information from the sensors and camera are available on a smartphone application and made accessible to the public. The smartphone app connects the drivers of vehicles to bicyclists and pedestrians along the corridor. Users of the app can receive warnings about traffic conditions directly around them. For instance, a cyclist riding on the side of the road who wants to change lanes or make a turn can receive a warning about oncoming vehicle traffic behind him or her which would cause a collision. In this way they are able to change their plans or wait until there is no traffic conflict. Similarly, pedestrians who want to cross the road at an intersection can receive warnings about vehicles which may be turning into the crosswalk from the perpendicular road and therefore advised to wait before crossing to avoid being struck. The smart corridor thus enhances safety of vulnerable populations such as pedestrians or cyclists who have little or no protection.

11.5 San Francisco

San Francisco, California, was one of the finalists in the (smartcitysf.com) US Department of Transportation smart city challenge. San Francisco is a large metropolitan city with several universities and technology companies.

San Francisco is known for its electric cable cars and in the recent past embarked on an electrification program. One strategy that San Francisco has adopted is to expand the number of electric vehicles on its roadway. The city has installed more than 100 electric vehicle charging stations and promotes the use of electric vehicles (EV) in the city.

Another smart city initiative is the provision of free Wi-Fi in major hubs in the downtown area. Several main roads have been outfitted with free Wi-Fi for use of the residents and guests.

In addition to new Wi-Fi, San Francisco instituted programs to replace conventional lights with LED lights which are much more energy efficient.

San Francisco launched its integrated mobility smartphone application which allows travelers to route trips, book trips, and pay for trips all in one interface. In addition, expansion of car-sharing services and promotion of bike usage and rentals have been key changes in the approach which city planners have used to make sure the city is a "smart city."

San Francisco has forged partnerships with institutions such as UC Berkeley to improve its data collection and data analytics of the smart city program.

Technology related to autonomous vehicles and connected vehicles is a major part of San Francisco's smart city initiative. The city has partnered with several private companies to build and expand the usage of autonomous vehicles and to implement the infrastructure needed for the growth of this program in the city.

11.6 Pittsburgh, PA

Pittsburgh, Pennsylvania, was one of seven finalists in the US Department of Transportation smart city challenge. Known as a city built by the steel industry, Pittsburgh has come a long way with major innovations and technology over recent years.

Pittsburgh is home to Carnegie Mellon university, as well as other state-operated colleges. Despite not winning the smart city challenge, Pittsburgh did embark on several smart city initiatives of their own.

One of the programs which were put in place was that of autonomous vehicles. Partnerships with private organizations such as Uber have resulted in the combination of autonomous vehicles with car sharing as a means of private transportation around the city of Pittsburgh.

Thus, residents of Pittsburgh can use their smartphone to call a ride-sharing provider such as Uber or Lyft and be serviced by a driverless vehicle which takes them to their destination.

Pittsburgh has been heavily in the field of robotics and the use of robots for automation of many tasks normally performed by humans.

Another strategy which smart city planners in Pittsburgh implemented was the use of adaptive traffic signals. The traffic signals in one part of the city are equipped with smart cameras and connected by a network to a central computer (pittsburghpa.gov).

When vehicles approach an intersection, video signals of the volume of traffic are sent to the computer processing the information and an algorithm adjusts the phase length of the red light. The roadway with the highest volume of traffic at that time will get serviced by a longer green light to clear the traffic and reduce the congestion from that intersection. Any emergency vehicles will be able to clear the roadway by accessing the infrastructure and the traffic signal controls.

11.7 Chicago, Illinois

Chicago is the third largest city in the United States by population. Chicago has made important steps to incorporate smart technology into its city planning. Installing hundreds of new LED lights into its streetlight network has allowed the city to save energy (ameresco.com) and millions of dollars each year on its electric bill.

Chicago has created an open data portal with over six hundred datasets with live data feeds from various sources such as law enforcement traffic signals which are video enabled and stream live pictures about the intersections and roadways in the city. This allows the emergency responders to make better decisions when someone calls for help since the operator can see live feeds about the conditions and activities in that area.

This database is available for developers who can use the information to develop a variety of applications for commercial or private use.

In order to support the data flow of information, the city has installed hundreds of sensors across a network which collects data on temperature, air quality, humidity, as well as traffic volume.

Chicago is notorious for parking congestion in the downtown area. With access to these datasets, new applications have been developed for parking. Drivers can use apps on their smartphone to locate and pay for parking spots anywhere in the city, which reduces their travel time and takes more vehicles off the road while reducing carbon emissions.

Chicago has partnered with several local universities to help automate their smart city development while also using technology which is sustainable and commercially available.

Along with the initiatives previously mentioned, Chicago also has dozens of smart digital-connected billboards prominently located along major interstates and sporting venues in the city.

11.8 San Jose, California

San Jose is a large metropolitan city in the northern section of California. With a population of over 1 million residents, it is the home to major technological corporations such as eBay, PayPal, and other well-known Silicon Valley companies.

San Jose has made major steps in the smart city planning of its urban environment. Developing partnerships with technology companies such as Intel, the city planners have created and installed a network of sensors across the city to measure and collect data about a variety of physical characteristics. Such information includes air quality, traffic flow, and noise pollution.

Once the data is collected, it is transmitted to the central data processing servers and subsequently to back offices for analysis. This analysis will provide insights into high-traffic areas, and recommendations are made as to what needs to be done to correct the situation. Similarly, knowledge about locations with high carbon emissions will allow the city planners to implement policies to combat that situation.

Other major projects in the San Jose smart city development include the Service Cloud (sanjoseca.gov), which was implemented with a partnership with Oracle corporation.

This application allows the residents of the city to access government services by smartphone and make requests which would previously be done manually.

Because of its Silicon Valley location, San Jose has ready access to many technology companies which can contribute to the smart city objective and implement solutions for collecting and processing data. Major universities nearby also form a basis for research and testing technologies such as autonomous vehicles and electric vehicles.

These partnerships serve as a catalyst for a sustainable smart city with new technology demonstration and implementation.

11.9 Tampa, FL

Tampa is a mid-sized city on the west coast of Florida. It is located approximately 80 miles to the west of Orlando and is situated next to the Gulf of Mexico.

Tampa is one of three sites (the other two are in New York and Wyoming), which was chosen to participate in a US Department of Transportation connected vehicle (CV) pilot program. In the Tampa study, over 1000 drivers were selected to retrofit their vehicles with connected vehicle technology which allowed the vehicle to communicate with the other vehicles on the road as well as with the installed roadway infrastructure and with pedestrians (theacvpilot.com).

The major roadway selected for the Tampa study is the Selmon Expressway, which is a reversible expressway located near downtown Tampa. During the morning rush hour, the entire road is dedicated to flow in one direction. In the evening rush hour, the traffic flows in the opposite direction to relieve congestion coming out from the downtown areas.

Due to its reversible nature, there is a high risk of wrong way traffic particularly from visitors who are unaware of the operation of this expressway.

With the installation of connected vehicle technology, the Expressway authority can now communicate the information to vehicles and warn them of the fact that they are driving in the wrong direction at that time of the day, thus avoiding a wrong-way collision. Similarly, on some portions of the roadway, there are sharp turns which end into a traffic signal. These areas have a high number of rear-end collisions. With the implementation of connected vehicle technology, this allows for vehicle to vehicle (V2V) communication and increases driver awareness if there are vehicles stopped at the traffic signal [2].

Because the drivers in oncoming traffic are now aware of the stopped vehicles, the risk of a rear-end collision is reduced and drivers approaching the traffic signal from behind can avoid a collision.

11.10 New York City

New York City is an advertising and technology intensive community with a lot of major corporations located in the tri-state region.

The city has embarked on several smart city initiatives [3] which are aimed at reducing congestion and improving the lives of its citizens.

One of these initiatives is the LinkNYC initiative (link.nyc) which is aimed at replacing the obsolete payphone which is located on many streets in the city and nearly every street corner in downtown NYC with a network of digital kiosks.

These kiosks provide free Wi-Fi service and a place for residents to charge their smartphone. It also provides connectivity to government services and information including traffic status. Residents can get near real-time updates on traffic conditions (including traffic accidents) nearby. Residents can also access 911 services from the LinkNYC kiosk.

The system provides nonstop digital advertising space where businesses can reach pedestrians and vehicular traffic. This advertising in turn pays for the operation of the kiosk and services.

New York has partnered with technology companies such as Qualcomm to provide the broadband network and service to the kiosk.

New York City has also been chosen to participate in the US Department of Transportation connected vehicle program. Along with cities of Tampa Bay, Florida, and the state of Wyoming, the connected vehicle pilot program focuses on regions which have high crash rates and seeks to use (V2I) and (V2V) technology to help reduce those crash rates and improve the flow of traffic.

11.11 Madison, Wisconsin

Madison is a medium-sized city in the state of Wisconsin. The city of Madison has a population of approximately 250,000 (cityofmadison.com) people with the metropolitan area about 800,000 population. Madison is the location of University of Wisconsin at Madison.

Madison has developed a smart city plan and continued to implement smart city initiatives. The city managers and planners structured their smart plan according to a three-step approach.

The Madison smart city plan includes a system for intelligent data collection and analysis.

This data collection is accomplished by a network of sensors which monitors air quality, vehicular traffic, and pedestrian traffic. The data is open and available to anyone such as for the development of applications for smartphones. The system also allows the traveler to plan and pay for trips on a single platform.

The city of Madison also implemented an autonomous vehicle and connected vehicle program to reduce carbon emissions and to improve safety through smart infrastructure.

Using dedicated short-range communication devices, which are installed at traffic signals throughout the city, connected vehicle in the city mass transit fleet of buses can communicate with each other and the infrastructure to prevent collisions.

Adaptive traffic signals at the intersections provide for smooth flow traffic. The data on the volume of traffic flowing through each intersection is processed and analyzed. Using sophisticated algorithms, the traffic signal timing is adjusted to match the volume of vehicle traffic present in the intersection at each point in time.

Smart infrastructure is the other pillar of the city of Madison smart city program. Thus, the city has installed a variety of physical characteristics such as temperature and air quality.

11.12 Reno Nevada

Reno Nevada is a mid-sized city in Nevada; with approximately 250,000 population, it is the third largest city in Nevada.

The city presented a plan for the US Department of Transportation's 2016 smart city challenge. The city has partnered with a number of university and technology firms to develop its smart city plan.

Some of these partners include the University of Nevada at Reno, Dell, and Intel.

The smart city initiatives which Reno focused on were autonomous vehicles (unr.edu), traffic signal sensors, and blockchain.

In partnerships with the University of Nevada at Reno, the city has developed smart city infrastructure to support autonomous vehicles. The intelligent mobility program implements deep learning algorithms in a testbed of vehicles and along the highway and allows for a fleet of self-driving electric vehicles. These vehicles include mass transit buses as well as private cars.

Using smart cameras and sensors installed on traffic signals, the city can monitor traffic flow and adjust the timing of the signals to reduce congestion and allow emergency vehicles to clear the road in the event of an emergency.

Blockchain is a major technology which is being implemented in the city. The research into using blockchain to verify data communications between the vehicle and the infrastructure has shown that it is possible to improve the safety of transmission of such data communications by preventing hackers from accessing or changing the data being communicated. As more autonomous vehicles take to the streets, this verification becomes vital to the safety of the public.

Blockchain LLC has purchased a plot of land near Reno Nevada, to develop a smart city built on blockchain technology. The concept of the blockchain smart city is that all transactions would take place with blockchain as the technology for processing information.

Blockchain LLC plans to focus on technologies of artificial intelligence, nanotechnology, and 3D printing.

11.13 Louisville, Kentucky

Louisville, Kentucky, is a major city in the western side of the state of Kentucky. Louisville has a population of over 700,000 residents making it the largest in the state. The well-known Kentucky Derby race is run in Louisville.

Louisville has created a plan for smart city improvements and has several smart city programs in place.

One of the initiatives is the widespread use of cameras and digital video of the metropolitan area to capture near real-time data. This data provides information about traffic volume and pedestrian counts along main roads in the city. This data is transmitted to the cloud database and analyzed for recommendations and improvements which government transportation planners can make to reduce congestion at peak commuter time periods at key locations in the city.

The city has also used the camera data to help fight crime by implementing intrusion detectors with the local law enforcement authorities. This allows for faster response and much more accurate location identification.

The smart city program management has also partnered with the Waze smartphone application (louisvilleky.gov) to help provide information to the city's residents. The application provides aggregate data of drivers at major locations across the city. On the other hand, the city provides information to the Waze app about construction and road closures which will affect drivers in those areas.

11.14 Richmond British Columbia

Richmond BC is a city in the western province of British Columbia, Canada. Richmond is neighboring Vancouver, BC. The city has embarked on a smart city program and has made progress implementing its smart city plans. Like several other cities its size (of about 200,000) people), it has deployed a network of IOT sensors across the city to collect data on parameters of interest.

The data collection procedure includes database information as well as cameras at signalized intersections. All the data is connected by and transmitted over an advanced fiber optic network to a data hub. After the analysis phase, the information and recommendations are then transmitted to the public for action and use.

Due to the diversity of the population of Richmond, the dissemination of information to the community is enhanced with technology and dependent on translation technology to reach all segments of the community. With over 70 percent of the residents from Asia (richmond.ca), who do not speak the native English, the information system must be able to be disseminated in multiple languages.

Richmond British Columbia was a finalist in the Canadian smart city challenge and has partnerships with multiple technology organizations to implement its smart city initiatives. Some of these partners are involved in the data collection phase and provide the fiber network for communications, while other partners provide analytic and data science services.

The Canadian smart city challenge provides funding for the finalists to begin the transformation of their city in the $10 million category.

One of the challenges which Richmond faces is the fact that it comprises several islands spread across a wide area. This requires greater communication and coordination for building the network communication infrastructure.

11.15 Toronto, Canada

Toronto, Canada, is the capital of Ontario, Canada. Toronto has a population of nearly 3 million people, which presents problems of traffic congestion for its residents.

Toronto has developed plans for its smart city initiative and partnered with Google's Sidewalk Labs (sidewalklabs.com) to help implement the technology.

For this project, the Quayside area along the waterfront of the city was selected as the testbed for the smart city implementation.

Several technology improvements have been proposed for this smart city project, some of which include smart cameras to monitor traffic flow and parking. These cameras would have their video signals transmitted to the cloud and analyzed for recommendations. The information can also be used to adjust the traffic signals to reduce the length of time that a vehicle waits at a red light if there is no cross-street traffic, or oncoming traffic which is turning into the intersection.

The smart city initiative also includes the use of sensors which can detect air pollution and monitor noise and temperature along the roadway.

Because of concerns about data privacy, the smart city is still under review and under consideration. Several privacy rights groups have brought up issues about how the data will be collected and used. Moreover, questions about data access have arisen.

The process of ensuring that the data is secure and that no individual's privacy right will be violated is ongoing.

The city of Toronto has also partnered with private industry companies such as Microsoft and QuadReal to implement their smart city project.

11.16 Boston, MA

Boston, Massachusetts, is a major city in New England. With a population of over 600,000 residents, it is the location of several education institutions such as Harvard University and Massachusetts Institute of Technology.

Boston was one of the cities involved in the US Department of Transportation smart city challenge. The city has continued to work on its smart city program and partnered with several private organizations and universities to implement its smart city initiatives.

Boston has taken a strategic path to develop its smart city program which includes smart cameras (boston.gov). The smart cameras can collect data on the number of vehicles on the roadway, as well as number of bicyclists and pedestrian counts. This information can be uploaded to the central database and, after analysis, used to provide recommendations for changes to the supply of city services to better meet demand. With the large concentration of students in the Boston and Cambridge area, bicyclists and pedestrians are particularly vulnerable, and data on how to improve safety is of paramount importance.

The city of Boston has also prioritized the use of bicycle sharing to help reduce the number of vehicles on the road and the carbon emissions and congestion that is involved with high volumes of vehicles.

New initiatives with car-sharing companies and autonomous vehicle providers such as Lyft have resulted in the testing of self-driving cars used for ride sharing in the city.

The use of sensors in the smart city program has numerous applications that help solve a variety of problems such as parking detection and reservations, air quality carbon emissions, and temperature as well as light monitoring. The new LED lights can be controlled automatically by measuring the ambient light in the vicinity and adjusting the light intensity as needed. This reduction in light usage also results in saving on the electric cost for the city.

11.17 Wyoming Connected Vehicles

The Wyoming connected vehicle program was included as part of a US Department of Transportation connected vehicle pilot program (https://www.its.dot.gov/pilots/). This pilot program had three locations with traffic conditions which could benefit from connected vehicle technology. These locations were:

- Tampa, Florida
- Wyoming (I-80)
- New York City, New York

The portion of Interstate 80 that runs through Wyoming consists of 402 miles of highway. I-80 is a major east-west corridor transporting over 30 million tons of freight each year.

A high percentage of the traffic on the section of highway which runs through Wyoming consists of large trucks and tractor-trailers. Wind speeds in excess of 10 miles per hour are common as much of the highway is above 5000 feet elevation above sea level. The roadway is mountainous and routinely covered with snow and ice during the winter months.

The combination of all these factors lead to a high risk of accidents and many fatalities.

The Wyoming Department of Transportation has implemented the connected vehicle program which was instituted by the US DOT and installed infrastructure to support vehicle to infrastructure (V2I) and vehicle to vehicle (V2V) technology along I-80. The connected vehicle program will include state vehicles such as police vehicles, snow plows, and trucks. Private vehicles will also participate in the program.

These vehicles are equipped with dedicated short-range communication (DSRC) equipment capable of transmitting and receiving signal information in the 5.9 MHz frequency spectrum range.

Over this band, several types of communication warnings are issued. These warnings include slow or stopped vehicles along the roadway, vehicles in distress calls, accidents which are miles ahead on the roadway, weather warnings such as snow or ice or construction zones which may require slowing down or changing routes [2].

It is important to note that with dedicated short-range communication (DSRC), the identity of the individual driver is not disclosed, and although the direction and speed information are available, it is used only for safety purposes. In the Wyoming connected vehicle pilot program, the windshield wipers of the test vehicles are equipped with sensors which activate when the driver turns the wipers on. This information is broadcast over the system so other vehicles can determine the presence of weather events such as snow or rain in that location.

DSWC has an effective range of about 900 feet or 300 meters.

On the Wyoming stretch of Interstate 80, the connected vehicle pilot program includes over 70 roadside towers for transmission and reception of signals and police vehicles, snow plow, vehicles as well as trucks which were retrofitted with DSRC equipment.

Several trucking companies which regularly use Interstate 80 have partnered with Wyoming Department of Transportation (WYDOT) to equip their private trucks with the necessary equipment as part of the pilot program.

The information collected on any section of the roadway about traffic conditions is transmitted to the Transportation Management Center (TMC) which is centrally located and then transmitted to the entire system so that all connected vehicles have access to traffic warnings about conditions along the corridor.

The Wyoming Transportation Management Center works with the connected vehicle infrastructure to inform residents about travel conditions along the I-80. In addition to the information from the connected vehicle pilot program, the Wyoming TMC receives information from a network of smart cameras which are monitored and transmit live video information to the TMC about road conditions which may be on or off the Interstate.

Wyoming Department of Transportation also has a mobile application (Wy511), which commuters can use to get information about road conditions in a near real-time mode. This mobile application can also be used with its global positioning service to determine and broadcast exact location. Therefore, if a motorist gets stuck or broken down and does not have access to any nearby road signs, the application can provide exact location information (within a few hundred feet) and transmit it to others who can help.

The Wyoming 511 mobile application also has a map function to provide road conditions on a visual display map of the state. The road conditions show the extent to which any event has impacted driving on a particular roadway. The information on the mobile application is also available (for the most part on the Wyoming state Department of Transportation website).

The Wyoming DOT has partnered with the University of Wyoming to provide training for state employees with the connected vehicle equipment installed on their fleet vehicles.

The connected vehicle program does always require an active driver in control as the vehicle is not autonomous. Other partnerships with the Wyoming DOT connected vehicle program provide assistance with data collection and processing.

Some examples of cloud services include Amazon web services and Cisco web services. These web services allow for the connection of vehicles to web applications for processing and to other web-connected devices.

11.18 New York City Connected Vehicle Program

The New York City connected vehicle program [3] is one of three locations chosen by the US Department of Transportation to participate in its connected vehicle pilot program (its.dot.gov). This program creates connected infrastructure along selected corridors in New York city. The connected corridors utilize dedicated short-range communication (DSRC) device electronics to communicate with vehicles as well as with the central web service where the data is processed.

New York City partnered with US DOT to define certain areas of New York City for its connected vehicle program.

The areas chosen for the connected vehicle program included:

1. FDR Highway along the east side of Manhattan
2. Flatbush Avenue in Brooklyn
3. Major north-south arterials in midtown Manhattan

Of the three sites selected by the USDOT for the connected vehicle pilot program, New York City will have the greatest number of vehicles participating with installed connected vehicle technology.

The vehicles equipped with connected vehicle equipment include mass transit buses, yellow cabs, third-party package delivery trucks and city vehicles.

Several applications have been designed and built for the New York City connected vehicle program; some of these include warning turning vehicles of pedestrians in crosswalk, red light running warning violations, and car stopped or disabled vehicle warning. In total there are 14 applications currently active which produce alerts of traffic conditions pertinent to the driving public.

The NYC connected vehicle program includes partnerships with several public and private organizations such as the NYC taxi and Limousine Service and Cambridge Systematics.

11.19 International Applications

The growth of smart city technology and applications has spread rapidly across the globe. The ideas, technologies, and partnerships have been implemented in cities such as London, Dubai, Seoul, and Singapore cities have implemented advanced infrastructure for collecting, processing, and analyzing data for improving their residents' lives in the areas of transportation, healthcare, and safety.

In the previous sections of this chapter, we covered several examples of smart city applications in cities across the United States. We will expand this coverage to include international locations as well.

Singapore is a city island nation located just south of Malaysia. The city has developed several smart city initiatives for improved living. One of these initiatives is the collection of data from taxis as they travel and mapping that location and speed data on a computerized system. This map helps to show the level of congestion at any part of the city by the concentration of the sensor data coming from that point.

Another smart city initiative involves the health and well-being monitoring system of elderly residents. The city has tested a pilot program which monitors activities of an elderly patient and records vital information about them such as time of wake time or sleep time. When the system detects an irregularity, it will send an alert to the caregiver(s) who can then check on the individual personally.

Sensors installed across the city provide information on the bus mass transit system, and that information is transmitted to commuters who use the data to make informed decisions about what route to take and when is the best time to take the bus.

Although Singapore is relatively small in comparison to other nations, it has taken significant strides toward smart city applications to their city.

11.20 Barcelona

Barcelona, Spain, has been one of the European cities which has developed a smart city program.

Barcelona smart city program has focused on a ground-up development whereby the citizens are involved with asking the questions about what problems they want solved by smart cities first. That is to say, before any sensors or smart cameras are installed, the citizens define a set of questions and issues which can be solved using smart city technology, and the appropriate steps are taken to install the infrastructure to solve that problem (barcelona.cat).

The next important approach the city has taken is the strategy that all the data is owned by the citizens, not the technology company. This approach ensures the security and privacy of the citizens' information. The data is used according to the will of the citizens of the city.

The Barcelona smart city initiative relies on sensors installed on major roadways to collect data on traffic flow and to measure volume data which can be useful in reducing congestion. In addition, low-cost sensors installed in individual homes can contribute to the data collection. Sensors in public measure air quality, temperature, and other environmental conditions. Barcelona is one of several cities to install garbage bin sensors which detect when a garbage bin is full and needs to be emptied. This saves time and cost because the collection is done just when it is needed and prevents full garbage cans from going uncollected, which improves the health of the citizens and livability of the city.

The introduction of smart sensors to monitor noise has also proven to help with noise pollution in the city center. As with other European cities, the usage of bike share and car share programs has been instrumental for reduction of fossil fuels and promotes green energy.

11.21 Dubai, United Arab Emirates

Dubai is a major city in the United Arab Emirates. Dubai has embraced smart city initiatives for more than a decade and has several ongoing initiatives.

Dubai embarked on the blockchain strategy, which is an initiative to use blockchain technology for all government services and transactions such as passports and licenses on the blockchain system. The city expects that it will be able to migrate all its transaction systems by the year 2021.

Another initiative which Dubai has developed is the paperless initiative. This program intends to create an all-digital platform for citizens to access government services.

Dubai residents will be able to pay bills and request government services such as business registration without filing a paper application. The Dubai government expects these changes will save the city significantly in man hours and materials over the next 5 years.

Dubai has placed the focus of its smart city initiative on its people, with its happiness initiative it expects to be able to measure the impact of any smart city changes which are implemented in the city.

Dubai has developed smart city initiatives for sectors of its economy. Some of these initiatives are based on [4]:

- Transportation efficiency and connectivity
- Energy grid
- Safety

Focusing on artificial intelligence, Dubai has created an AI lab which examines how AI can be used in combination with sensors and Internet of Things (IoT) to provide data and analysis to improve efficiency in these areas in its city.

In response to the transportation and energy initiatives, the city has a program in which they provide citizens with financial incentives to purchase electric vehicles. The city has also installed numerous electric vehicle (EV) charging stations to increase the use of electric vehicles and reduce the carbon emissions.

Dubai has also embarked on solar power initiatives, and residents have installed solar cells on their private homes to generate electricity which can be fed back onto the power grid. This will reduce the use of city electric power during peak consumption periods and the home user's electric bill.

The smart city initiatives which Dubai has embarked on include artificial intelligence (AI) applied to the three focus areas previously mentioned.

Dubai has installed a network of smart cameras along major intersections in the city which monitor the traffic performance through live video feeds as well as with traffic volume counts. This traffic information is fed back to a central location where transportation agency managers can analyze the data and provide transportation planning for major events and initiate what-if simulation modeling to predict how the traffic system will behave under given conditions.

Other Implementations

The application of X-Road in Finland is particularly interesting, as the country has successfully applied it to its e-Residency project. The X-Road (x-road.global) is an innovative tool that allows for a more interdependent system among governments and for more seamless transactions between countries. Cross-border services are growing in popularity, and now Finland has taken several steps toward making them more relevant.

Card payments account for a larger percentage of total transactions in the Nordic countries compared to other European countries. A report from the European Central Bank indicates that 47 percent of all transactions in the EU are made with cards. The two Nordic states were also ranked at the top when it comes to mobile payments, with an even greater percentage than Britain and Germany.

According to comparative surveys, countries like Estonia and Finland lead the way in e-invoicing. In 2013, 3.2% of invoices were sent electronically, but by 2016, that number had jumped to over 40%. That's double the amount of e-invoicing in just 3 years!

The nation of Estonia is known for using NFC technology as one part of its m-wallet initiative. It began by allowing its citizens to pay for public transit with their mobile phones and has since expanded its reach to include paying for parking via the same method. Increasingly, the average citizen uses his or her mobile phone to pay in stores; m-wallet makes it possible to receive a receipt on a mobile phone in real time against mobile payment.

References

1. Beyond traffic: The smart city challenge. https://www.its.dot.gov/factsheets/pdf/SmartCities.pdf
2. Harding, J., Powell, G. R., Yoon, R., Fikentscher, J., Doyle, C., Sade, D., Lukuc, M., Simons, J., & Wang, J. (2014, August). *Vehicle-to-vehicle communications: Readiness of V2V technology for application* (Report No. DOT HS 812 014). National Highway Traffic Safety Administration.
3. New York City. New York Connected Vehicle Pilot Deployment Program. https://www.its.dot.gov/factsheets/pdf/NYCCVPliot_Factsheet.pdf
4. Happiness Agenda. Retrieved on September 26, 2019, from www.smartdubai.ae

Chapter 12
Summary

12.1 Introduction

In this summary we will review the topics covered in the earlier chapters in the book related to smart cities.

We have examined smart cities from several angles and reviewed several implementations of smart city technology and applications. From a technical standpoint, we have shown through several examples how smart cities utilize technology to help develop their strategy for serving their residents.

For instance, we have shown how San Diego used smart streetlights to collect data on traffic volumes and to save electricity on expensive conventional lights.

Other cities such as Columbus, Ohio, implemented autonomous shuttles with electric vehicle (EV) technology and have planned connected vehicles dedicated roads for the demonstration of this technology. Further, the US Department of Transportation set up three domestic locations for a connected vehicles pilot program to explore the use of this technology. The three locations chosen for this pilot program are:

1. Tampa, FL
2. Interstate 80, Wyoming
3. New York City

We covered these locations in earlier chapters showing how the connected vehicle system was implemented at each location.

Data collected by the sensors and smart cameras are transmitted via communications networks to a central location for processing. This is often a "cloud" technology architecture. The system for storing and processing data serves as a portal for several functions related to smart city operation. The system must be able to perform the tasks of:

© The Author(s), under exclusive license to Springer Nature Switzerland AG 2023

D. Graham, *Integrated Electronic Payment Technologies for Smart Cities*,
https://doi.org/10.1007/978-3-031-38222-2_12

- Administration: That must be able to control the devices on the network. For instance, turn on/turn off devices.
- Big Data Analytics.
- Ability to develop applications using the data collected.

12.2 Big Data Analytics [1]

The common theme that has emerged throughout the earlier chapters is that one must have a unified strategy when planning a smart city. Every decision is interconnected. The type of sensors that you use to collect your data depends on what type of measurement and analytics you want to do. The type of analytics you want to do depends on what type of problems you have in your urban location.

We have mentioned data analytic methods in some previous chapters. Here, we will focus on analytics which is somewhat more customized for smart city applications and smart city transportation problems.

Some typical transportation problems faced by smart city planners might include: How can we time the traffic signals in an intersection or set of consecutive intersections so as to maximize traffic flow and minimize travel time? Another problem which transportation planners might need to solve is: Where should the mass transit bus stop be located to be accessible to the maximum number of riders while minimizing risk of pedestrian accidents?

These and similar types of questions can be answered using data collected from sensors and cameras in a smart city network. The data from smart city sensors and cameras is often very sizable due to the sheer number of devices connected to the network and the fact that they are transmitting data on a continuous basis.

Over the past few decades, there has been a tremendous growth in computing power and storage capacity which allows for computers which are able store these huge amounts of (big) data. Complexity in CPU power has also increased so that computers can perform complex instructions and calculations in a fraction of a second which would previously take much longer.

Correspondingly, data science techniques have been developed to provide useful insights from such large quantities of data. Here we will introduce some of these techniques which can be applied to smart city optimization.

12.3 Mathematical Optimization

Mathematical optimization includes both linear and nonlinear optimization methods. The optimization may be either deterministic or stochastic. These methods can be useful in smart city applications which require solutions of problems related to maximizing or minimizing mathematical expressions [3]. One could use this technique to minimize travel time through a smart city network or to find the shortest distance through a street grid from an origin to the destination.

12.3.1 Decision Trees

Decision trees are a data science technique which can be used to make decisions involving multiple choices. A decision tree can be used in a smart city problem which involves making decisions about the best location for a mass transit station as an example.

Decision trees can be used with continuous variables or categorical variables. Decision trees can be subdivided into classification trees (CARTS) or regression trees.

12.3.2 Forecasting

Forecasting is a useful data science technique which has long been used to predict values of variables of interest. Such variables can be related to smart city transportation networks. The peak hour demand for mass transit bus or subway transportation can be predicted using forecasting techniques with the historical data collected from the network.

Forecasting is a broad area of mathematical analysis which includes regression models. Linear regression techniques have been applied to smart city data which is collected using sensors and transmitted to the cloud. The Big Data analytics from these external sources can provide real-time information about trends related to the surrounding environment. Thus, for example, the light sensor on a streetlight can indicate a darkening environment due to cloudy weather, and the system can then turn on the streetlight to provide light for the commuters and pedestrians nearby [1].

Simple regression models will consist of an explanatory (independent) variable which "explains" the resulting value of the (dependent) response variable. The model is said to be a good fit when the coefficient multiplying the independent variable is chosen such that the resulting regression equation closely predicts (estimates) the actual observed value of the response variable.

Regression models can include both continuous numerical data and categorical data, and any regression model with several independent variables is called a multiple regression.

In many real-world problems, there are multiple variable features which must be known in order to predict the outcome response variable. Thus, for example, an image of an object can be identifiable by a finite number of "features" or variables, and a machine learning algorithm can classify any object based on whether it possesses the same features.

Regression models which do not meet the conditions for linearity are often fit using a nonlinear expression.

12.3.3 Logistic Regression

Logistic regression is a data science technique which is often used by smart city planners to predict the probability of certain events occurring. Logistic regression models can be used in transportation to detect the probability of an accident occurring at certain intersections given certain criteria. Using historical data from smart cameras and sensors, we can train the model on the conditions that have led to accidents in the past and select those variables which are likely to contribute to accidents at that location in the future.

12.3.4 Random Forest

Random forests are used in data science to classify and predict the outcome response given specific input variables. Random forest can be used with numerical variables or with categorical variables. Random forests will generate several decision trees based on some algorithmic principles and come to a solution for each tree. In the event the problem is numeric, the random forest will use an average as its solution. Random forests can be used to obtain answers to questions in a smart city with a high level of predictive accuracy.

Smart city data which is collected by sensors and smart grid networks is often stored in open-source form for use by the city planners and administrators. Developers may also use this data to create applications for additional insights into the operation of the city. Many applications developed from the smart city data utilize these data science techniques mentioned above as well as others. We will provide more examples of the usage of these data science methods to smart city planning in the upcoming sections of this chapter. In most cases the data that is collected from millions of sensors and IOT devices is aggregated, and individuals remain anonymous. This is important for privacy and security.

12.3.5 Support Vector Machines

These techniques are useful tools for smart city planning. Like decision trees, support vector machines can be used to classify outcomes based on given independent input variables.

Support vector machine attempts to find the vector which most accurately defines the boundary between two or more classes.

12.4 Internet of Things

Throughout the book we have mentioned the term "Internet of Things" (IoT) several times. In the context of the smart city, IoT is a very important technology. We will delve a little deeper into the structure of the IoT architecture and explanation here.

The Internet of Things (IoT) loosely refers to sensors, smart cameras, and any other device or object which can be connected to and controlled through a communication network. The data collection devices can be connected to each other and a gateway on the network. The gateway will transmit the data from these devices to the cloud using network communication protocols. Different communication protocols are used depending on the level of control desired.

Some of the protocols used include MQTT, XMPP, and HTTP. The data transmitted to the cloud will typically report the state of the connected device, i.e., on or off. The protocols will also allow communication from the administrator to change the status and send other commands depending on the device. Thus, for example, a light can be set to on, off, or dim (change in intensity).

The cloud data can first be sent through the gateway which connects to the smart devices and sensors. The gateway collects the data and aggregates them. In some architectures the data can be transmitted to the cloud directly instead of through the gateway.

Sensors connected to the network are in communication with the cloud servers and are perpetually transmitting data and receiving instructions. The data from the sensors is real-time data which can be analyzed and acted on by the stakeholders.

The scale of the Internet of things has exceeded even the number of computers connected to the internet. Sensors, radiofrequency identification (RFID) tags, and other Internet connected devices have been estimated to exceed billions over the next few years.

This large quantity of diverse devices connected always results in many terabytes (and in some cases petabytes) of data to be stored and analyzed by the smart city.

12.5 IoT Protocols

Big Data thus requires special transmission protocols to ensure that the information and messages between the sensors and the cloud servers are delivered efficiently.

The different protocols operate in different bandwidths on the frequency spectrum and have different transmission ranges. The network planner must select the protocol which is best suited for their network characteristics and topology.

We have mentioned some of the protocols earlier in the chapter, such as MQTT (message queueing telemetry transport), and we expand the discussion to introduce protocols such as constrained application protocol (CoAP) and WebSocket.

The different network protocols can be compared based on their efficiency as measured by the packet size and energy usage. Some experiments will show that the efficiency of MQTT and CoAP protocols can be affected by packet sizes in transmission.

The smart city sensors are low-power devices which are battery operated. The sensors can continuously send data on the network for 5–10 years depending on battery life.

12.6 IoT Applications [2]

Internet of Things applications are numerous and span literally every industry. The connection of a sensor object to a network transforms the device from a product to a service. This is a fundamental transformation for the business and consumer alike.

This basic transformation in the use of the product requires new business models to take advantage of the new functionalities which IoT brings to the customer.

As mentioned, Internet of Things (IoT) applications include connected vehicles (CV), and the information and sensing capabilities are included with that technology.

Connected vehicles can retrieve information about the vehicle from the onboard diagnostics (OBD) port which is typically installed in late model cars near the front (dashboard) area of the vehicle.

The onboard diagnostics port records information about the location, speed, and direction of the vehicle.

This information can be transmitted to other vehicles and to the roadway infrastructure to help prevent accidents and improve travel decisions, when there is congestion or dangerous conditions on the route.

With several automobile manufacturers, intelligent sensing mechanisms are built into the vehicle which can sense the mechanical state of the engine, as well as the condition of peripheral objects such as the tires and the lights. With this information sent to the central processing location, diagnostics can be performed, and in the event repairs are necessary, this information can be transmitted back to the vehicle which can then be taken for preventative maintenance.

12.7 Connected Vehicles

In an earlier chapter, we briefly discussed the connected vehicle pilot program implemented by the US Department of Transportation in some locations in the United States. We will expand on that discussion in this summary as we believe that connected vehicle technology will be a prominent feature in vehicles of the future.

The US Department of Transportation pilot program for connected vehicles began in 2016 in the conceptual phase. The system was deployed in March 2018 in Tampa, Florida. The Tampa program included 1000 vehicles, 10 mass transit buses, and 8 streetcars which were equipped with connected vehicle technology.

The Tampa in-car vehicle technology consisted of a standalone unit which is placed in the trunk or near the front of the vehicle. It is not connected to the OBD port as some connected vehicle (CV) units are.

The unit reports the speed, acceleration location, and heading ten times per second using dedicated short-range communication (DSRC) protocol.

Similarly, the roadway is equipped with the technology to send and receive messages using dedicated short-range communication (DSRC) to the vehicles which are connected to the network.

These roadway units are located at traffic signals at the intersections every few miles apart. The units transmit signal phasing and timing (SPAT) messages and speed warnings for vehicles which are exceeding the speed limit or moving too fast in highly congested areas.

The roadside equipment and in-vehicle units are provided by vendors which have partnered with the city of Tampa on the deployment of the connected vehicles.

12.8 Connected Vehicle Pilot Measurement and Evaluation

The connected vehicle (CV) pilot program was deployed in Tampa in 2018. After deployment, the next phase of the project is the measurement and evaluation phase. The method used to evaluate the (CV) program in Tampa can be adopted and with modification applied to other connected vehicle programs in other locations.

Here, we discuss the major component sections of evaluation. The first sub-area or component of the Tampa evaluation is safety.

12.8.1 Safety

The connected vehicle program was evaluated on the criteria of whether it resulted in improved safety. The safety measures are similar to how autonomous vehicles are evaluated and include such items as reduced crash rates, reduced conflicts (especially at intersections), and general safety improvements for pedestrians.

Along with the safety criteria, the next component of the evaluation is mobility (Achilleas, K. (2019, October 14), personal interview).

12.8.2 Mobility

Mobility improvement due to the connected vehicles is of major interest to commuters and stakeholders. The mobility is measured by reduced travel times from origin to destination, reduced overall congestion, and mass transit flow and throughput in congested intersections. In the Tampa pilot program, the mass transit buses were given priority at the traffic signals to allow for the maximum number of commuters to move through congested areas.

The next component of evaluation is the agency efficiency.

12.8.3 Agency Efficiency

The agency efficiency refers to the ability of the transit agency to get the message out to the general commuting public. Information about wrong-way drivers on the expressway must be sent to the other vehicles in the network to avoid collisions. In addition, general information about crashes and high congestion must be passed on to the drivers who may be affected in that area so that they can choose an alternate route.

The fourth component in this evaluation scheme is the environment.

12.8.4 Environment

Environment is improved by less congestion and less delay in the network, since when there is less congestion, there are fewer cars idling at intersections on the system.

With fewer cars delayed, there is less emission of noxious gasses from vehicles. The level of emissions is measured at certain locations and regular time intervals to assess the improvement of this factor.

In order to facilitate the evaluation of this connected vehicle program, the planners set up six use cases to measure the performance metrics they chose. The six use cases are as follows:

The first use case is wrong-way driving at Twigg Street location. The second is the level of traffic backups and congestion at a specific curve on Twigg Street. The third use case is the bus priority at traffic signals. The fourth use case relates to the streetcar warning to oncoming and cross traffic in the city. The fifth use case relates to pedestrian warnings for the intersections and oncoming traffic. The sixth use case refers to the heavy pedestrian traffic at the courthouse location.

The city of Tampa has partnered with several original equipment manufacturers to provide equipment and services for the connected vehicle pilot program. The roadside infrastructure units (RSU) are provided through the partnership with Siemens corporation. The in-vehicle (onboard) units are produced by several different manufacturers; however, the ones used by the connected vehicle pilot program in Tampa include Sirius XM and Commsignia.

The data collected by the V2V and V2I equipment is uploaded to a central database on the ITS (Intelligent Transportation System) datahub which houses information from a wide variety of sources (source: https://www.its.dot.gov/data/).

The data on this database is open source and freely available to the public and can be accessed for development purposes.

The ITS datahub also stores data on the Wyoming connected vehicle pilot program as well as the New York City connected vehicle program. These datasets can be used for analyses of the CV program and evaluation of improvements due to the implementation of connected vehicle technology.

12.8.5 Security

As with any program which collects data on the public, there is a concern with the privacy and security of this data. In addition, questions about how the data will be used must be answered.

It is important to note that for the connected vehicle program, several steps to ensure security have been taken. These security provisions include: No individual can be tracked using the equipment. The data is encrypted before transmission across the network. The vehicles are assigned a rolling identification code which changes every few minutes. The system ensures that there are no hacker messages transmitted on the network.

These security and privacy concerns are addressed by the protocols and hardware installed.

12.8.6 Privacy and Security

The issue of privacy and security goes beyond just the connected vehicle pilot in Tampa.

As mentioned earlier in the book, Toronto and other cities planning smart city implementation have raised concerns about this.

Whenever data is being collected on the public, the rights of the individual to remain private and not have their privacy intruded on is of serious concern. Many smart city planners maintain that the data collected is not individual data but that the data is aggregated so that no one individual is identifiable or tracked. Another safeguard is that the data is anonymized and images from smart cameras only contain counts of individuals.

Privacy rights advocates argue that even when data is anonymized and aggregated, the behavior patterns of groups of individuals can be used for gain by private for-profit companies. As we saw in the case of some of the smart cities mentioned earlier, the issue of who owns the data collected is of major concern as we move into the era of smart cities worldwide.

Another major issue in the smart city debate is that of security. Whether a vehicle is used as an autonomous vehicle or a connected vehicle, there are certain vulnerabilities to which the technology can be susceptible.

In particular, with regard to an autonomous vehicle, some of the security questions being asked by the public are as follows: What would happen if there was a software glitch? How does the vehicle respond when there is a hardware failure? Other issues are more malicious, such as how does the vehicle software defend against outside hacking into the system?

Similarly, for connected vehicles, although there is always a human in control of the vehicle, security concerns arise if the connected vehicles receive an erroneous message due to security lapses on the network.

12.9 Blockchain

In earlier chapters in the book, we discussed briefly the concept of Blockchain and its impact on smart city payment. Blockchain technology has been proposed as a method to safely send and receive communications over digital networks.

The connected vehicle utilizes low-power networks to send and receive messages over a short-range using DSRC (dedicated short-range communication) protocol. Blockchain technology has been explored as a way of improving the safety of the connected vehicle.

The decentralized structure of blockchain can make it difficult for spurious messages to be transmitted through the connected vehicle network. Information such as speed of the vehicle, location, and heading can be sent to the web and stored on a blockchain. Because this information is distributed throughout the network, it cannot be tampered with without other users being aware of a false message.

In smart city transportation, we can use blockchain to enable the transaction payment systems.

In many smart cities, the transportation services are made available on a web application. Therefore, the different modes of transportation such as mass transit bus, car share (Uber, Lyft), subway or light rail, and bike sharing can be accessed on a single integrated platform for route planning purposes.

Although each mode of transportation may charge a different fee for their service, the integrated payment for a single trip can be done via encrypted blockchain technology.

The blockchain also has capability for smart contracts. The smart contracts can be used to eliminate the need for a third-party verifier in a transaction between two entities.

The smart contract is coded into the blockchain to help ensure that the service requested is completed according to the predefined agreed-upon terms. With the smart contract on the blockchain, the funds, for any product or service, are only released at the appropriate intervals. Insurance companies for example can use a blockchain to distribute funds when the work on a claim has reached certain preset milestones.

Blockchain technology has applications to supply chain networks. The blockchain can enable tracking of products from their origin to the consumer with accuracy and speed that is not possible today. Experiments with blockchain tracking have shown that a single product can be identified and tracked to its source in less than a minute with blockchain, whereas the same product from the same blockchain would take up to a week with current methodologies.

As we know, natural disasters strike in the form of earthquakes, hurricanes, tornados, and even wildfires from time to time across the globe. When that occurs, the individuals and corporations reach out to help those who were affected by the disaster. This usually happens with the help of a third-party charity who collects the donations for the affected countries or locations. Often in these situations, people are contacted by scammers who fraudulently pretend to be representing those in need. Blockchain technology can help ensure that the funds collected go to the persons affected while eliminating the need for a third party. This reduces the incidence of fraud.

12.10 Blockchain Technology Applications

Blockchain technology has also been applied to reduce voter fraud in governmental elections.

Several blockchain-based web applications have been developed for different industries. Auger is a forecasting blockchain-based technology that is distributed across the network and generates input from multiple sources.

Another blockchain-based application is Ripple, which is useful for trading across the network without the need for a central third party and speeds up transaction processes across international boundaries.

12.11 Cloud Services

With the onset of big data and the associated science of big data analytics, there came the need to store huge volumes of information and create actionable insight from it for smart city planners to use.

Several private industry companies have developed "cloud" services which can facilitate the storage and management of these petabytes of data.

Some of the companies offering cloud computing services are:

1. Amazon Web Service
2. Microsoft Azure
3. Google Cloud

These companies provide a shared information technology infrastructure which other companies can access on demand as needed.

In this cloud service environment, individual companies do not have to buy and maintain their own IT hardware infrastructure. In addition, it is easy to scale to a larger level when the need arises.

On the cloud, several services are offered to companies and individual users.

Software as a Service is a method of placing commonly used software which have traditionally been installed on individual computers or company networks and migrating them to a centralized cloud system. Some software which has been migrated and uses cloud services are Microsoft 365, Google applications, and CRM software such as Salesforce. These are accessible from any location or platform.

Platform as a service is a method of providing access to a development environment on the cloud. This PaaS provides operating systems and execution environments for developers.

12.12 Amazon Web Services

This cloud computing service provider is considered by many as one of the largest such providers in operation. It is estimated that 50–80% of all fortune 500 companies have used Amazon Web Service (AWS) [4] for cloud computing.

Some major companies which use Amazon Web Services include Airbnb, Netflix, and GE, among others.

Amazon Web Services has data centers worldwide so that each data center has an estimated 300,000–500,000 servers [2].

One of the features of Amazon Web servers is the ease of use. AWS provides on-demand server capabilities for individuals as well as major corporations. Cloud computing provides Infrastructure as a Service which makes servers available to institutions. It also provides Platforms as a Service (PaaS) and Software as a Service (SaaS).

For the smart city connected vehicle, the cloud can serve as a repository for the huge amount of data collected by the Internet of Things (IoT) devices.

Amazon Web services provides storage solutions, database services, as well as analytics services which are applicable for smart city operations.

Amazon Web Service possesses redundancy built into its cloud infrastructure. AWS data centers are divided into regions and availability centers which are structured for easy transition from one server to another in the event of change of traffic.

AWS provides blockchain services for transaction and integrated payment. AWS also provides IoT to store and analyze data from smart devices.

There are other cloud computing providers which provide similar services to Amazon Web services. Microsoft Azure is considered a major cloud computing provider based on market share. Other cloud service providers such as Google cloud platform, IBM cloud, and VMWare are major cloud computing-based companies based on market share.

AWS was first deployed in 2006 and was one of the first companies to provide cloud computing on a pay per use basis.

Microsoft Azure was deployed in 2010 and provides cloud computing services from over 40 data centers throughout the world. Cloud computing services such as machine learning, Internet of Things, Analytics, content delivery, networking, DevOps, mobile, and integration are among those offered to Microsoft Azure users. Like the other major cloud service providers, Microsoft Azure has hundreds of services offered across broad categories such as database, storage, and analytics.

Microsoft Azure is very useful for application development and storage, as well as application performance monitoring (www.azure.microsoft.com).

Google Cloud is a cloud service provider which was launched several years ago. Google cloud services offer an array of options such as machine learning, big data services, computing services, storage services and networking services. Google has made available to other users some of its own infrastructure which it uses for its products such as email (www.cloud.google.com). Some of the major companies using Google cloud services include Twitter, PayPal, and eBay.

Cloud services can be obtained with a variety of options. Public cloud consists of servers which can host multiple applications and data from multiple sources and organizations.

Private cloud consists of servers which are dedicated to a single organization for hosting and access of its information and files.

References

1. Dash, S., Shakyawar, S. K., Sharma, M., et al. (2019). Big data in healthcare: Management, analysis and future prospects. *Journal of Big Data, 6*, 54. https://doi.org/10.1186/s40537-019-0217-0
2. Salman, T., & Jain, R. (2017, March). A survey of protocols and standards for internet of things. *Advanced Computing and Communications, 1*(1).
3. Graham, D. (2013). *A comparative evaluation of FDSA, GA, and SA nonlinear programming algorithms and development of System-Optimal methodology for dynamic pricing on I-95 Express* (PhD dissertation). University of Central Florida.
4. Explore AWS Platform, cloud products and capabilities. Retrieved from the world wide web www.aws.amazon.com, October 8, 2019.

Index

© The Editor(s) (if applicable) and The Author(s), under exclusive license to Springer Nature Switzerland AG 2023

D. Graham, *Integrated Electronic Payment Technologies for Smart Cities*, https://doi.org/10.1007/978-3-031-38222-2

MIX
Papier aus verantwortungsvollen Quellen
Paper from responsible sources
FSC® C105338

If you have any concerns about our products,
you can contact us on
ProductSafety@springernature.com

In case Publisher is established outside the EU,
the EU authorized representative is:
Springer Nature Customer Service Center GmbH
Europaplatz 3, 69115 Heidelberg, Germany

Printed by Libri Plureos GmbH
in Hamburg, Germany